Mastering Site Reliability Engineering in Enterprise

A Complete Guide to Resilient Systems & Chaos Engineering

Florian Hoeppner
Francesco Sbaraglia

Apress®

Mastering Site Reliability Engineering in Enterprise: A Complete Guide to Resilient Systems & Chaos Engineering

Florian Hoeppner
Tech Strategy Senior Manager
Accenture
New York, NY, USA

Francesco Sbaraglia
Chief Technology Architect Site Reliability
Engineering & AIOps
Accenture
Sydney, NSW, Australia

ISBN-13 (pbk): 979-8-8688-1447-1
https://doi.org/10.1007/979-8-8688-1448-8

ISBN-13 (electronic): 979-8-8688-1448-8

Managing Director, Apress Media LLC: Welmoed Spahr
Acquisitions Editor: Aditee Mirashi
Desk Editor: James Markham
Editorial Project Manager: Jacob Shmulewitz

Cover image designed by Pixabay

Distributed to the book trade worldwide by Springer Science+Business Media New York, 1 New York Plaza, New York, NY 10004. Phone 1-800-SPRINGER, fax (201) 348-4505, e-mail orders-ny@springer-sbm.com, or visit www.springeronline.com. Apress Media, LLC is a Delaware LLC and the sole member (owner) is Springer Science + Business Media Finance Inc (SSBM Finance Inc). SSBM Finance Inc is a **Delaware** corporation.

For information on translations, please e-mail booktranslations@springernature.com; for reprint, paperback, or audio rights, please e-mail bookpermissions@springernature.com.

Apress titles may be purchased in bulk for academic, corporate, or promotional use. eBook versions and licenses are also available for most titles. For more information, reference our Print and eBook Bulk Sales web page at http://www.apress.com/bulk-sales.

Any source code or other supplementary material referenced by the author in this book is available to readers on GitHub. For more detailed information, please visit https://www.apress.com/gp/services/source-code.

If disposing of this product, please recycle the paper

To our families, whose patience, love, and unwavering support made this journey possible, and to the global SRE family shaping the reliable digital future.

Table of Contents

About the Authors

Florian Hoeppner is a seasoned professional technology strategist and advisor for tech operating models. He is an enterprise SRE subject matter expert and distinguished engineer with a deep understanding of tech operating models and transformations. Florian is passionate about tech strategy, combined build-run teams, and optimizing tech operations, and he has spoken and published extensively on these topics.

He is the creator of the EngineeringOps radar, a yearly publication showing tech engineering and operational capabilities. He holds a degree in Media Information Systems and a Master of Science in Digital Media. Florian currently lives in New York and shares his knowledge online in multiple publications with strategic and practical insights into Tech Operating Models, SRE, Chaos Engineering, and Tech Engineering practices and AI enterprise reinvention solutions. He has published the book *Competition as Motivation* with AV Akademikerverlag.

Francesco Sbaraglia is a distinguished site reliability engineer (SRE) and a recognized expert in the field of Chaos Engineering and DevOps. With an extensive career spanning over two decades, Francesco has garnered a wealth of hands-on experience as a practitioner and innovator, establishing a profound mastery of cutting-edge AIOps technologies and methodologies.

In addition to his technical prowess, Francesco has distinguished himself as an accomplished author, contributing numerous insightful tech articles and authoritative books across a spectrum of subjects surrounding SRE, Chaos Engineering, and

DevOps. Francesco is also a public speaker, sharing his insights and best practices in SRE, observability, and chaos engineering at renowned industry conferences, such as SRECon, SREDays, and DevOpsCon. He is passionate about combining systems engineering principles with observability tools to ensure seamless operations and improve software engineering practices.

About the Technical Reviewers

Eric Jager is the author of *Getting Started with Enterprise Architecture* and a certified master architect in the field of enterprise architecture. He is also a certified TOGAF enterprise architecture practitioner, certified business architect, and ArchiMate practitioner. He is familiar with various architecture methodologies, including the TOGAF Standard and the Zachman Framework.

Eric has been practicing enterprise architecture for over 15 years and is considered to be a thought leader in the field of enterprise architecture for his extensive knowledge and experience in the development and application of enterprise architecture. He easily leverages architecture to translate business strategy into implementation.

Eric has worked for many organizations ranging from government agencies to healthcare institutions and lectures on architecture at Eindhoven University of Technology. He also speaks at various conferences and seminars.

Eric likes to focus on the practical and pragmatic application of enterprise architecture and writes about his experiences on his blog.

Saaniya Chugh is a senior technology consultant with over a decade of experience in IT service management, enterprise automation, and ServiceNow platform implementations. She has led large-scale digital transformation initiatives across a range of clients and industries, specializing in AI-driven solutions, IT governance, and workflow optimization. Based in Canada, Saaniya works at the forefront of intelligent IT systems, advising clients on the effective implementation of the ServiceNow platform.

A recognized thought leader in the ServiceNow ecosystem, she regularly contributes to industry forums and supports the ServiceNow community. She has been recognized at multiple platforms for her expertise and continues to influence the evolution of modern IT operations and enterprise service platforms.

About the Co-Authors

Yvette Zzauer is a leader in organizational change, with a track record of driving large-scale transformation across global enterprises. She has led enterprise-wide technology transformation initiatives, including a flagship program that influenced behavior change for over 16,000 employees worldwide. Additionally, Yvette specializes in designing and implementing future-ready operating models that align with strategic business goals across diverse industries. Her approach is both systemic and human-centered, ensuring that transformation efforts are sustainable and embraced at every level of the organization. Based in New York, Yvette holds two master's degrees and brings a unique blend of academic rigor and practical insight to complex change challenges.

Michele Dodic is a leader in site reliability engineering architecture and transformation, enabling large enterprises to integrate observability into complex, hybrid environments. In addition to architecting end-to-end observability frameworks, Michele is an active member of several SRE and DevOps communities. His focus is on promoting SRE principles, practices, and culture to help organizations improve reliability and operational efficiency. He also shares his insights as a speaker at industry conferences, including All Day DevOps, ObservabilityCon, DevOpsCon, SREDays, and others.

Acknowledgments

This book is the product of countless voices, ideas, and inspirations drawn from across disciplines, industries, and generations of SRE and Chaos Engineering practitioners. I've long lost track of the number of training sessions, articles, books, conversations, and public talks that shaped my thinking before and during the writing of these pages. Their influence is woven throughout the book—explicitly in the references and implicitly in the perspective it brings.

It is through these diverse cross-domain insights that we can finally break free from the constraints of traditional IT operations and DevOps, evolving toward a more resilient, adaptive model rooted in Site Reliability Engineering and Chaos Engineering. There is deep value in learning not just from success but from failure—from missteps, disruptions, and breakthroughs, even when they come from worlds far beyond computing.

A special thanks to Yvette and Michele for their invaluable insights, thoughtful critique, and the knowledgeable, inspirational minds they brought into the writing process. Their contributions—whether in shaping ideas, challenging assumptions, or helping articulate complex topics—greatly enriched several chapters and pushed this book toward greater clarity and depth. Your unwavering support, perspective, and belief in this journey have meant more than words can express.

Yvette Zzauer is a leader in organizational change, with a track record of driving large-scale transformation across global enterprises. She has led enterprise-wide technology transformation initiatives, including a flagship program that influenced behavior change for over 16,000 employees worldwide. Additionally, Yvette specializes in designing and implementing future-ready operating models that align with strategic business goals across diverse industries. Her approach is both systemic and human-centered, ensuring that transformation efforts are sustainable and embraced at every level of the organization. Based in New York, Yvette holds two master's degrees and brings a unique blend of academic rigor and practical insight to complex change challenges.

ACKNOWLEDGMENTS

Michele Dodic is a leader in site reliability engineering architecture and transformation, enabling large enterprises to integrate observability into complex, hybrid environments. In addition to architecting end-to-end observability frameworks, Michele is an active member of several SRE and DevOps communities. His focus is on promoting SRE principles, practices, and culture to help organizations improve reliability and operational efficiency. He also shares his insights as a speaker at industry conferences, including All Day DevOps, ObservabilityCon, DevOpsCon, SREDays, and others.

We would like to thank Mirco Hering for his inspiration and for constantly pushing our boundaries, which made it possible to bring our SRE experience to paper. A heartfelt thank you to Marco Torre, Sonny Dewfall and Joe Galley for their great help in building the SRE community and for always bringing fresh ideas that advanced the global SRE practice. Special thanks to **Brian Berg**, a remarkable technology and strategy leader, whose vision, mentorship, and sharp thinking inspired every step of bringing this book to life on paper. His influence not only helped crystallize our message but also shaped a sensational introduction section that set the perfect tone for what follows.

Our deepest gratitude to our Editor, whose patience, sharp editorial eye, and insightful pushes helped refine and elevate the manuscript in all the right moments. Thanks also to our production editor and technical reviewers for going the extra mile to follow our thinking, embrace our unorthodox structures, and support a writing style that breaks the mold. At the end of the day, we are hands-on, charismatic, inspiring techny under the skin—and this book reflects that spirit. A heartfelt thank-you to Apress for believing in this vision and providing the platform to transform our knowledge into a book, making it accessible to professionals and enthusiasts around the world.

We believe this book will serve as a meaningful resource for the next generation of software engineers, platform engineers, DevOps professionals, SREs, and chaos engineers. Our hope is that it not only informs but also inspires—encouraging readers to challenge conventions, think across domains, and engineer systems that are more resilient, adaptive, and human-aware. This book reflects the personal views and opinions of the authors. It is not written on behalf of, nor does it represent the official views, policies, or positions of the authors' employers, including Accenture. Any mention of companies, vendors, products, or organizations is for descriptive purposes only and should not be interpreted as endorsement. You can find more insights and ongoing discussions on our blog at `www.nextgen-sre.com` —follow us there to stay connected.

—Florian and Francesco

Introduction

The systems we build today are no longer just tools—they are ecosystems of complexity, resilience, and relentless change. Traditional operations models are straining under the weight of this evolution, and even DevOps, once revolutionary, is showing its limitations in the face of dynamic, distributed, and failure-prone environments.

In today's dynamic digital landscape, IT is no longer merely a cost center—it's the very heartbeat of enterprise success. Organizations are navigating a new era where speed, reliability, and adaptability have become not just strategic goals but fundamental survival traits. Our customers expect rapid innovation, flawless execution, and continuous improvement, mirroring the effortless experiences delivered by consumer apps and digital platforms.

Site reliability engineering (SRE) is an innovative practice designed to meet these heightened demands. Born at the intersection of software engineering and operations, SRE is the essential strategy for organizations striving to build robust, scalable, and future-ready systems.

We journey through the foundations of SRE, exploring why traditional approaches fall short and how a new operational model is required—one where reliability is embedded deeply into organizational culture. Drawing from real-world experiences and proven methodologies, we'll guide you through the key pillars of SRE: Observability, Error Budgets, Blameless Postmortems, Chaos Engineering, and more. You'll discover how leading organizations are overcoming common hurdles, such as legacy infrastructures, regulatory complexities, and deeply entrenched mindsets.

Highlighted through vivid analogies, such as the dramatic imagery of speeding down an autobahn without seatbelts or guardrails, we bring attention to the crucial balance between innovation and risk management. We illustrate how the developers of today have emerged as central figures within enterprises, transforming traditional hierarchies and redefining responsibilities and accountability.

At its heart, this book emphasizes a holistic view of resilience, integrating advanced practices from SRE into the broader operational context of enterprises. It is about more than just technology—it's about fostering collaboration, driving cultural shifts, and leveraging tools and automation to deliver tangible business outcomes. Our approach, termed "Enterprise SRE Next Gen," pushes beyond basic concepts to reshape the entire enterprise landscape, empowering build-run teams and redefining traditional operational boundaries.

Whether you are an engineer, an IT leader, or a business stakeholder, you will find insights to help your organization thrive. Prepare yourself for a journey of transformation—one that turns complexity into competitive advantage and positions your organization to succeed not just today, but in the rapidly evolving future of technology. Additionally, the rapid advancement of AI, generative AI, and agentic AI is revolutionizing SRE by reducing on-call fatigue, accelerating issue diagnostics, and enabling proactive disruptive event prediction through historical data insights, transforming how organizations anticipate and respond to operational challenges. This book distills decades of practical experience onto paper, aiming to make tomorrow's systems and organizations better, stronger, and more resilient than today's.

## CHAPTER 1

Introduction to Site Reliability Engineering

Welcome to the start of an incredible journey, one built on real lessons learned, pitfalls faced, practical advisory, and thought leadership that shape the disciplines of site reliability engineering and chaos engineering. This book is your guide through the principles, practices, and mindset shifts that make these fields so impactful.

In this chapter, we will explore the journey of site reliability engineering (SRE). We will begin by examining its necessity in the current landscape, delving into the challenges that organizations must face. Next, we will provide an overview of SRE's history and a fundamental introduction to its core principles and essential components. Finally, we will discuss its immense potential and advantages, emphasizing the significant value it brings to businesses.

Why SRE?

IT organizations are no longer cost centers. They now make up the fabric of the entire enterprise, and their mission is speed.

The Need for Speed

Something in IT has changed in the last 10 years. We are now focused on speed. Our customers are used to getting new features in short iterations, and the business department requests high speed from their IT delivery. Developers must frequently deploy in production, one deployment after another, in short iterations.

© Florian Hoeppner, Francesco Sbaraglia 2025
F. Hoeppner and F. Sbaraglia, *Mastering Site Reliability Engineering in Enterprise*,
https://doi.org/10.1007/979-8-8688-1448-8_1

IT leadership is guiding their teams on a high-speed route. The car is now on the autobahn, and there are no speed limits anymore. What do managers feel when they drive down the autobahn at 120 miles per hour? Do they think about the speed and how much it has accelerated in the past years? Do they think about guardrails? About seatbelts? Do they touch their seatbelts just for a second to see if they are still there? Is it still working? I do.

Let's talk about seatbelts and guardrails. Because what has changed is that now we are all driving down that autobahn. If you haven't got there yet, you will. Everyone will go there, sooner or later. But currently, the guardrails are missing. Nor are there seatbelts. There's just the speed pedal and the traffic sign that says no speed limit.

The current business strategy requires IT to accelerate small and fast changes. It requires small innovations and quick pilots to test and try out new ideas. This is the new way of interacting with customers. Customers are demanding fast improvements and ongoing changes. This is what they know from their App Store. My sports app has been improving constantly, and my car gets regular updates and improves its capabilities. This is what we expect and what we are used to.

Employers expect fast, regular updates from their internal IT. People have become used to having reliable access to data and functional systems. And recently, personal dependence on IT has increased dramatically. The reputation of an internal IT department or an entire brand depends heavily on reliability. These expectations peaked during COVID. When we all worked remotely, our dependency on IT skyrocketed. We no longer had the option to do business in person. In this new reality, stability and reliability are unquestionably as fundamental as the air we breathe.

Software Development Has Taken Center Stage

There are more changes in IT besides the thirst for speed. In this new world, developers have become central. How does this happen? The current focus is now on the consumer. The customers consume our company's product, and most of them rely on IT services. The banking client buys shares with our banking app. For families, the autopilot is the trusted guide on their vacations. And the Metaverse is pushing the IT experience to the next level. In many companies, new hires have their onboarding in virtual reality.

IT determines how our customers experience the whole company, from products to services. This gives developers a central business role. IT services and software development have now taken center stage. This change is irrevocable. The tide has turned; we will never go back.

While inevitable, this change isn't occurring without pushback. Some businesspeople resist the momentum of IT taking over the company. While continuing to make demands on IT departments, they often resist partnering. You can see that by the frequent hesitation to fill the role of a product owner.

Some businesspeople don't realize that IT has already taken over. Some others are searching for their new place. We see some businesspeople signing up for programming classes. We see them financing shadow IT and know that they are running their servers under the table.

Finally, with no-code and low-code, the business itself has become a citizen of the IT department. This shift marked a turning point: gravity within companies no longer pulls toward the top of the business hierarchy; it pulls toward the developers. Developers became the new queens and kings of the IT kingdom. Now, a new force is emerging: vibe coding, very intelligent prompt engineering. By blending natural language with advanced AI tooling, vibe coding allows both developers and non-developers to harness software creation at unprecedented speed and scale. It's not just about writing less code; it's about thinking differently. In this new paradigm, influence doesn't just lie in knowing how to code but in knowing how to ask, precisely, strategically, creatively. The future belongs to those who can command machines with clarity and intent. This shift pushes business and IT closer together, making development a shared responsibility and accelerating innovation cycles. However, it also demands new reliability practices and governance to maintain trust as AI-driven solutions evolve rapidly and unpredictably.

Keeping Applications Up and Running

More change can be seen in how money is spent in IT. In this area, two souls live in our breasts: building and operations. Most companies spend too much on operations and too little on engineering. For each dollar companies spend on engineering, they pay double or more in operations.

Each area is dependent on and connected to the other. Changes in production and human intervention in operations are the leading causes for the growing number of incidents. Missing automation in operations leads to more manual work and causes even more disruptions in production systems. As complexity increases, so does the time to find errors and provide solutions.

In traditional companies, the pain caused by incidents is poorly distributed. The symptoms lie with the customer and the production support. The cause lies with the developers. The engineer who created the code does not feel the outages. They don't get notices when their code causes an interruption. Developers do not even see the budget spent on operations. The budget is discussed only on a leadership level. Engineers usually don't know the spending for their application in operation, the number of incidents, or the service requests. Most do not even know the names of the operations experts.

Imagine if the developer team were accountable for the operational spend. They would have to buy the operations service or even spend their own time doing the manual operation tasks. They would have to fix incidents and purchase cloud storage. Even better, imagine they would also experience the positive effects of their application. Imagine they could see the rising numbers of customers loving their work and the positive resonance a feature has on the customers. And finally, imagine they would even get a part of the earnings from the use of the application. What would this world look like?

Rethinking How to Organize the Creation and Maintenance of Software

Resilience has become a competitive advantage for companies. Companies that exceed the required regulatory requirements stand out. Imagine a car manufacturer making a commercial pointing out that its autopilot has fewer accidents than its competitor's system. Or imagine a trading platform communicating that its application uptime is the highest in the industry.

Summarizing what we just described:

- IT processes are speeding up more and more but lack safety guidelines. We are driving our car fast, but without seatbelts. We've sped up our deployments, but we aren't considering additional stability improvements.

- The developers are becoming queens and kings. IT is the core of every product and customer experience. It determines the reputation of a firm, and it will do so even more with the upcoming Metaverse, vibe-coding, or anything that comes next.

- Spending on operations is out of control and increasingly does not address the tasks that are actually causing the costs.

Creatively responding to this situation requires business and IT to take a step back and think holistically. By "holistically," we emphasize the necessity of including multiple departments in our transformation.

What we are doing impacts how we handle **finances**. When we consider employee contracts, we must include **human resources**. We will revisit **compliance** and **security** requirements, improve the quality of **data**, include **architecture**, and change beliefs in **engineering** and the **organizational structure**. In fact, we are thinking about everything.

We start by increasing **operational empathy**, which means that developers will learn what it means to do operations. We want developers to understand the impact of their work. They should also understand the tasks and the thinking of people responsible for the production system and the feedback they receive. In this way, we raise the consciousness of non-operations workers about what it means to do operations.

Including all these pieces is crucial to improving resiliency and stability. Operational people get a direct, unfiltered response from clients who are using our product right now. Operational empathy leads to customer empathy. We close the circle between people who are changing the application, people who are maintaining and operating the application, and users of our application.

Resilience is the single most crucial element in the future of enterprises. Our businesses depend on IT systems, which are now the main product. IT determines how the customer experiences and thinks about our products and, ultimately, our company. At the same time, when IT changes at increasing speeds, increased pressure is placed on our systems. Stability must be introduced. Where does it come from? From developers to those who change the application.

The idea of IT resilience is not new. For many, it's closely tied to disaster recovery. These are plans that are made in case a catastrophe happens. After the plans are made, they are revisited and tested once a year. For many, this process is not taken seriously. It's just checking a box, and the plans are rarely improved.

Disaster recovery activities are executed after something happens. What we need is the ability to see before something happens. We need a sixth sense, like a spider sense. With the speed we work, transform, and consume, after-the-fact is too slow.

The resiliency we require must be built into our IT processes, applications, and tools. It must also be built into the mindset of our people and the structure of our organization. We navigate uncertainty by sensing and observing. Resiliency means reacting precisely and quickly when necessary and constantly implementing improvements.

Google recognized the importance of resilience early on. While Google Search is still one of their flagship products, the spotlight is steadily shifting toward Gemini, shaping up to be the next-generation gateway for finding information, getting things done, and interacting with AI more naturally. It's hard to imagine a world where products like Search or Gemini go offline, where people suddenly can't find what they need or are forced to turn to a competitor. Google combined multiple IT capabilities under the term site reliability engineering (SRE). Ben Treynor Sloss founded the first SRE team in 2003. Following in Google's footsteps, other companies started to employ site reliability engineers. In the early days, it spread first to other web-based companies. Later in 2016, the first book on SRE was published. Now SRE is becoming mainstream.

At its core, SRE seeks resiliency by breaking the silo between operations and development. It applies engineering principles to operations and focuses relentlessly on the customer's needs and experience.

Our team has been applying the Google principles to leading enterprises with a focus on a highly regulated industry. Early in our journey, we experienced a strong pushback. We realized that what worked for Google doesn't necessarily work for companies born before the cloud. These companies still use a mainframe, have a hierarchical leadership culture, and must follow strict regulatory requirements. Also, these traditional companies have been in a highly competitive situation for centuries. They are under high cost pressure because they do not have the quasi-monopoly that tech firms enjoy. When you are not born in the cloud, you will find it challenging to implement the same practices as Google.

The second aspect we noticed is that SRE is heavily discussed and defined on a team level. Books are written for software developers, production support operators, and monitoring specialists, all of whom work on a team level. The concept of a department, the IT organization, and the enterprise is lacking or entirely missing. However, this is precisely what is needed when we want to apply SRE to a traditional company with a legacy.

Here, we share our approach, which has been applied and tested.

When we work with an enterprise and enable it to use the advantages of SRE, we can easily see an immediate positive impact. Applying engineering practices and mindsets to operations allows us to move from reacting to outages to preventing them.

After years of focusing on ITIL, we've realized that SRE is the first movement that comes from operations, which influences all the other parts of an IT organization. We call our approach **Enterprise Tech Resilience**. Companies now require resiliency from the onset. With COVID and wars starting in Europe, your business needs to be resilient. Your business department will not ask for site reliability engineering. However, SRE provides the stability and resilience it desperately needs.

Handling the Unexpected

We started the chapter by focusing on IT because our primary audience is IT organizations. But let's also consider the business side.

On October 4, 2021, at 15:39 UTC, Facebook, Instagram, WhatsApp, and Oculus became globally unavailable for more than six hours. Not only this, company employees could not enter the building or conference rooms. It was reported that Facebook lost around $60 million in advertising revenue. It also lost some of its reputation.

This was not the only thing that happened on October 4. When people could not log into Facebook and all its other products, they started increasingly using other apps with similar services, such as Twitter, Telegram, Signal, Gmail, and TikTok. This change in behavior caused slower experiences across these substitute applications when users were searching for similar services to mock and complain about Facebook.

The outage and its financial impact illustrate the pressing demand to rethink resilience, and the behavior chain underlines the need to revisit it from end to end. Even if you were not directly impacted, your SaaS vendor was. You may have experienced a source code repository that is no longer reachable. No one could deploy code in production for some hours. You may have had other experiences related to your vendors. It's likely that at some point, you were impacted by this.

This kind of outage changed everything. Now, when global SaaS providers fail, the New York Times and the Washington Post are affected. Not only do IT people know this, but your business stakeholders are also informed. Your businesspeople realize that their business relies heavily on IT resilience. Now, IT resilience is on the CEO's table, and demanding questions are being asked.

The COVID-19 pandemic was the tipping point. Employers on your business team who cannot work because of an outage, reputation loss, or customer attrition have a direct financial impact, making resilience the number-one topic. During the pandemic, SRE became mainstream. IT resilience became a key topic in traditional companies, not only for functionality but also as an external requirement from regulators.

Rethinking Operations

The time to talk about guardrails and seatbelts is now, when you are pushing down that spread pedal. On a micro level, we know that development teams' work is becoming increasingly complex. All indicators of higher complexity are increasing: the number of microservices, the unsolved technical depth, the manual tasks a team must perform to keep everything alive, and the number of lines of code. Systems are becoming more complex. A single person cannot oversee an application anymore.

Managing Increased Dependencies

On a macro level, we see that operations responsibility is shifting. The adoption of the public cloud and the move to standard SaaS and PaaS providers are shifting responsibility. On-premises and cloud are now coexisting. Teams working in a waterfall style and highly advanced build-run teams also coexist. An enterprise must consider all types and characteristics in its policies, access controls, entitlement structure, and compliance discussions.

The Demand for Stability and Reliability

What applies to our external customers also applies to our internal customers, our peers, and colleagues who use our internal applications. Remote work has irrevocably changed the requirements for internal applications. But even if all the employees are called back into the office, the world will never be the same again.

Our internal customers are no longer accepting slow systems and the wrong data. We have seen people leaving the company because internal systems blocked them from executing their work as efficiently as they could. When we fight for the best talents on the job market, our internal systems must accept this challenge and perform where and when required. Companies must go above and beyond when they want to keep their best people.

Now, with smartphones, all people have easier access to applications. A traditional computer is no longer required anymore. Everybody compares the experience they have in their private lives with the knowledge they have in business situations. When my private travel booking application works better than the application I must use in my business situation, I get mad. The applications I use on my mobile for free are optimized for usability, the graphical user interfaces are sophisticated, and the user workflows are considered and make sense. The applications work when I need them, and I don't have to wait.

When I use business applications, I often do not even understand what I have to do. After I click a button, I must wait till the application reacts to see if the application is available at all. The experience in my private life increases the requirements for stability in business situations. Users want to have a better understanding, and they know that similar software is doing it better.

History and What Is SRE

Leading companies have already found ways to be highly reliable. Many capabilities combine under the term "SRE." Most of them aim to give developers the confidence to deploy changes into production quickly. These developers are confident, but at the same time, they know that not all is perfect. They assume that errors will happen. They plan for it. Planning means to have enough wiggle room. The practice of SRE provides the confidence that whatever happens, we are prepared for it. It makes the system stable against human errors.

Leading companies, not only those born in the cloud, have increased transparency on their critical user journeys. These companies sense and observe errors, latency, data quality, and more to quickly understand if they must improve the stability of their systems. They also predict errors and upcoming instability.

One key aspect of this transformation is that it also enhances human lives and increases well-being. It gives knowledge workers time off when they are out of the office. By empowering, SRE reduces personal stress and decreases obligation.

Let's say we start by collecting data for work on the weekend. When do teams release their software? On Friday or Saturday? Let's change that. We want to give quiet weekends to our people by adding more options to their tool belt for making releases. Advanced release strategies let us deploy new features into production without interrupting the user, and they offer a safety net.

But SRE can do more for the well-being of your workforce. We can focus on optimizing incident management by reducing noise and improving transparency and automation. We reduce noise by validating alerts. Not all alerts lead to an incident that is so critical that our staff must solve it on the weekend.

We make relevant details on incidents transparent. Not all incidents on a weekend require our staff to start bug fixing on their laptops. We show relevant information on their mobiles.

Scripts that help solve incidents use automation. However, scripts must not be started from laptops. We offer the option to start predefined scripts from mobile devices. We want to provide options for investigating and solving incidents, keeping in mind the circumstances of our company and our people.

These companies are treating the operational stability of software systems with the same care and importance as they treat the development of new features. The mindset of engineering helps to improve operations. Operations people are becoming engineers. They are changing the perspective from building "systems" to building "products." Development and operations work together as one team. Build-run teams bring many improvements to production, such as reliability and faster deployments.

A critical user journey doesn't stop at the end of an application. It includes multiple systems and requires observation both end-to-end and across systems. This keeps the customer experience at the heart of all development and operational activities. All team members are responsible for the product. Each person in the team has operational empathy and is proud of a successful product.

We also see that leading companies are testing their product boundaries. They simulate, test, and challenge their systems using methods that traditional product owners would have never allowed. These tests are demanding, but they know that only these kinds of tests will lead to tangible improvements.

There is, for example, chaos testing. Capabilities like chaos testing are embedded in the broader system. We do not just start with an advanced practice like that. To get to the point where chaos tests can be conducted, the teams and systems have to undergo multiple steps. These steps are not only technical. Mindset, process, and technical aspects all go hand in hand. We start with the belief that our software is broken. Nothing is 100% perfect in IT. We also embed chaos testing in our IT process landscape, for example, by connecting it with postmortems. We can take our existing shortcomings as the basis for writing the test cases. In addition, we first want our team to be good at observing their systems. They must have already applied the proper techniques and be mature enough to conduct this kind of test.

By working together with enterprises on their SRE transformation, we have realized that resilient companies share some core beliefs:

- Metrics and data are the fundamentals.

- Trust and improve the system; don't rely on a single person.

- Invest in preventing, not in reacting.

- Build a culture of ongoing change; all and everything is changeable.

- With each change, reduce the complexity; strive for simplicity.

- Improve collaboration; share your knowledge always, anytime.

Overcoming Key Challenges

All transformations involve multiple hurdles. To improve operations and to accomplish high reliability with speed, we must overcome key challenges. But some are really special to SRE, and we want to share them here.

There's No One to Hire: All companies are searching for SRE specialists, but there is no one in the job market. The job description combines skills from development, IT service management, and the desire to work in a challenging environment. No one applies, and these people are not easy to find.

Everybody Is an SRE: At the same time, many companies just renamed their operations teams to "SRE." Many people say they are SREs, but they are not. They don't have the required skills we are looking for. These skills will not lead to the desired outcome of our journey; they will just slow it down.

Business Involvement: At the start of the SRE transformation, practitioners often don't realize that buy-in from business is essential. From afar, SRE looks like an IT improvement. But it's not. Ultimately, we are improving for the customer. The customer is represented in an enterprise by the business department and has its representation in agile teams in the role of the product owner. When we design and create new features, we take, for example, the customer's forgiveness into account. We give it a name like "application latency" and convert it into numbers. The numbers let us consolidate the experience in a final error budget. This provides us with the possibility to balance innovation with reliability by listening to the experience our customers have. All that can only happen if we work closely aligned with our business department and our customers.

Legacy Infrastructure, Software, and Mindset: The companies we work with were not born in the cloud. Most are still using the mainframe. If it were easy to get rid of it, they would have done it years ago. Over decades, they've optimized their operations processes and established their own beliefs. Here, production support is not an area where you experiment. Change happens slowly and must begin by first winning hearts and minds.

In our transformation approach, we must plan accordingly so we know the impediments. We tailor our approach to the unique requirements of a demanding job market. The speed of the program and the design of the different steps that we'll share with you later take that into account.

Introducing Enterprise SRE 2.0

SRE is best described in the sentence, "It's what happens when engineers are doing operations." To define SRE, we first want to explore the existing silos and points of interaction in IT; what we call the four walls: business, development, operations, and infrastructure.

We use the concept of SRE to interact with, target, and break down each of these four walls:

We break down the barrier between **business stakeholders and operations** by providing transparency on critical customer use cases through the Error Budget and Service Level Agreements (SLAs).

We break down the wall between **developers and operations** by prioritizing resilience in the backlog and identifying relevant resiliency design patterns, such as those used after incident outages.

With the build-run concept, we entirely break down the wall between **operations and development**. It lets developers and operations work together as one team. Developers temporarily rotate to operations, and both share responsibility. There's no difference between the teams anymore.

We have two types of **infrastructure**: on-prem and cloud. To achieve a higher degree of reliability and increase speed, we rely on a product and platform structure. Everything that is not a customer-facing product is defined as a platform (maybe you know them as "horizontals"). On platforms, we build services to reduce the waiting time for product teams.

Another important fact is the scalability of the infrastructure. SRE highly depends on the possibility of scaling up and down on short notice. That way, infrastructure is not a hurdle to reliability. This is why we need to be in the cloud.

SRE happens in these four walls. It brings together business, developers, operations, and infrastructure. After we apply our advanced approach to SRE, Enterprise SRE 2.0, none of these walls will be intact.

To change established mindsets, we must go further. With Enterprise SRE 2.0, we are now able to tear the walls down by swinging the most enormous hammer: the **budget**. We want the development squad to be responsible for profit and loss (P&L). A team consisting of a product owner, developers, and operations must pay for their new features, operations, and infrastructure. The key to a successful transformation is including budgeting. We want product teams to be transparent about both their consumption and their financial impact.

SRE Cosmos

We must restructure the entire IT organization with SRE in mind. This is what we call the SRE Cosmos.

Compared to other technology movements, SRE is relatively new. However, different concepts, such as Agile, DevOps, and Cloud-Native, are well established, with their own culture, mindset, tools, and beliefs. Together, these concepts form a new narrative for enterprise technology. Each concept adds its chapter to our software lifecycle story.

The story SRE tells is outstanding on its own because it is a logical continuation of already existing concepts. In its simplest form, SRE is about applying cooperation between IT teams to break down silos. However, SRE is even more valuable than other technology movements.

We have a unique perspective on how SRE plays out in the enterprise and becomes embedded in its culture:

Our product-aligned teams deliver in short iterations with quick feedback from the customer on new features. The value of high reliability, a core concept of SRE, becomes apparent when our teams are working in fast and short iterations. When we apply **Agile,** we break down the wall between business and operations. We have a product owner in our squads who uses the error budget, a means of assigning business value to reliability work, to coordinate cooperation between business and IT. Decisions within the team can now be made by balancing reliability with functionality.

Using our transformation approach, SRE is a logical extension of **DevOps**. With DevOps, we make delivering software into production easier by automating our deployment debt (our CI/CD pipeline) and shifting the responsibility for testing from the quality team to the developer team. We also automate the manual tests.

With SRE, we automate our operational debt (including release engineering, which is often part of the CI/CD pipeline). We set up build/run teams, effectively transferring ownership of reliability and operability to the same "development" team. With build/run teams, developers create systems that are easier to operate and observe right from the start.

Combining SRE with **Cloud-Native** thinking improves many aspects of our software, including scalability, reliability, speed, and data transparency. With Cloud-Native, we shift left infrastructure work and improve reliability. Applying SRE principles to Cloud-Native development accelerates the migration to the cloud by providing easy access to observability data and ensuring recently migrated applications are robust by measuring SLIs and SLOs.

On-prem applications are often more difficult to modernize. They require ongoing cooperation between application and infrastructure teams. Application teams are also frequently asked to become experts in observability, as they are required to maintain their observability tools and data.

This is what we call the SRE Cosmos. A cosmos means viewing the enterprise as a complex and orderly system. We often must impose our order to harness the enterprise's complexity. But our SRE Cosmos goes beyond just SRE. We encompass and enrich the entire culture. From this perspective, traditional SRE is bound and limited.

Many organizations have started applying concepts like DevOps, Cloud, and Agility and are still on that journey. Most have gone through an Agile transformation, some are doing great work in DevOps, and others are focusing on Cloud-Native. Some value streams have been identified and have moved to a product and platform structure. Some organizations have considered combining all these aspects into a homogeneous whole, but that is rarely executed or even fully documented.

We cut through the noise to define the SRE Cosmos. We are extending the culture of SRE by looking into Agile, DevOps, and Cloud-Native. If we want to understand SRE and improve, we must see the complexity of its surroundings and explore how it all works together. Culture is complex, just like the SRE Cosmos, but it is also ordered. When we see it all together, structured, we understand the dynamic. All movements strive to accelerate the work in IT; everything centers around the customer, and developers must be empowered.

Optimal Working Model for SRE

Now, as we have defined the four walls of SRE and the SRE Cosmos, we want first to describe the optimal working model for SRE. Practitioners in our advanced model rotate between three general activities:

- Implementation

- Solving incidents

- Prevention

The entire team spends around one-third of its time on each activity.

Implementation and improvements require the core skill of software development to improve the application, build self-services, and automate manual tasks. Solving incidents is traditional operations work. It seeks to understand the root cause, find workarounds, and directly support the customer. Preventing downtimes includes observing the environment, learning from the past, and testing the boundaries. Here, we require new thinking to anticipate and optimize operations.

We know that almost no one has all three of these skills. However, we are not forming the optimal person; we find and build teams. Modern engineering is a team sport. This is what we call full-stack teams. In a rotation model, we let people rotate from one activity to another. We want them to be generalists in their team, but each of them has a focus. Over time, they will learn the skills they lack from their teammates. But also, they will know that they must rely on each other and cooperate.

Three Options for Placing the SRE Capability

Establishing capabilities at varying levels of the portfolio allows a balance between business process specialization and the depth of technical discipline. At the same time, it aligns with demand. In most enterprises, we see all three options used at the same time. This relates to the maturity of the development teams. When a development team is advanced in SRE, they have different requirements than a team just starting with it.

Option 1: SRE As a Full-Time Role in a Squad

This is at least one named, dedicated SRE in every squad. The typical characteristic and benefit of this is that SRE objectives have a single point of accountability at the practitioner level. It's easier to hold a balance between development, operations, and prevention. The dedicated SRE focuses on growth and improvements. SRE defines how the squad's products are to be operated at their peak, reliably and efficiently. The SRE reports to a center of excellence and the team. SREs in a full-time role have their SRE backlog in the same backlog as the squads.

Option 2: SRE As a Practice/a Skill for Everyone

If prioritized against features, anyone and everyone is skilled in SRE and executes SRE practices. We can name SRE leads or champions to ensure accountability and drive the community. It's a shared responsibility to define how products are operated and improve their reliability.

Option 3: Have a Centralized SRE Team

The centralized SRE staff is aligned with the transient squad. Most of the teams are not product-specific but tech-stack-specific. They accomplish task automation and reliability improvements in partnership with squads. Mostly, we see them as nonoperational. They are highly mobile across products and squads. They have two backlogs: one is a product backlog, and the other a central SRE COE backlog. This is a good starting concept when SRE teams are built up and not so many people are identified as SREs. So we are all together, and they can learn from each other and improve faster.

SRE Is Hard to Find

All the companies we are interacting with are facing the same problem. They must start an SRE practice from scratch. But this is not entirely true. In all companies, we see that people are already doing the job. Sometimes they are labeled differently, and mostly they do only half of the SRE job. Our first step is always to identify these people. We want to understand what they are doing and how they do it. Are they more aligned with development work or operations? Do they do more application work or infrastructure? Cloud or on-prem? These are the main characteristics we need to see in order to find the gravity of SREs in the company.

To start the practice, we need to understand what skills are required. We want SREs to be responsible for increasing the velocity of successful production change execution. They improve the efficiency of infrastructure and operations by optimizing architecture, tooling, and automation. Part of the skillset is understanding measurements to improve security, performance, maintainability, scalability, and usability.

The working model of SREs is highly interactive. SREs are consulting developers in architecture for reliability and scalability. They collect and use data to communicate and improve constantly with the user in mind, continuously evolving. They are engineers who have applied their craft to infrastructure and operations problems. SREs improve and build software to help operations teams and optimize systems to be scalable and highly reliable. They break down problems, simplify, and find convincing solutions. They value the improvement of daily work more than the daily work itself.

The outcome of their work is a highly reliable product after code deployment. Better operability means a better life for operation teams. The right balance of speed versus safety, the balance between new features and the stability of the system to optimize the customer experience. Balance between development, prevention, and operations. They help to create ultra-scalable and highly reliable software systems to support business functions.

A detailed job description depends on the technology stack being used. However, we must also remember that we will not find a perfect fit. The skill set we are seeking must have been developed over time.

We've had success searching for and upskilling people who are skilled at monitoring and observability. We also find DevOps engineers who understand how to automate and build a deployment pipeline. The third option is performance engineers, who know how to improve the performance of applications or infrastructure and are thus well-positioned to take on SRE work.

Hiring from outside adds another complexity. Many people in the market call themselves SREs, but their job description is pure production support. Many companies have renamed their production support workforce SRE because it sounded better, and they thought they needed SREs. Now, they have people calling themselves SREs, but without an upskilling program, they are far away from the described craft.

This renaming of operations people as SREs creates a problem for those who take the concept seriously. Most operations people lack the core software engineering skills. Some can make minor improvements in addition to their work in operations, but not to the extent that we can call it engineering. Software development is the most complex skill to learn.

The Potential and the Promise of SRE

After all the complications of implementing SRE, why should we do it? The results SRE delivers have a direct impact on our customers.

With the right release strategy, we reduce the number of incidents after changes and can easily roll back or roll forward when an incident occurs. Less manual work and more automation allow us to act faster and more reliably when an incident occurs.

Because of our measurement and optimized observability, we detect problems earlier. Our SLI/SLOs indicate when something is slowing down, when the customer already feels that the system is no longer optimal.

The resilience design patterns are proven solutions that make our systems highly reliable. Operating with build-run teams, the operational empathy of all team members allows us to learn faster about operational issues, which helps reduce the number of incidents.

Teams with a culture of mutual respect, who don't blame each other and thus feel more psychological safety, are faster. They are quicker to identify improvements in postmortems, ask for help, and change ways of working. We see teams with low psychological safety in a "yes, but…" reflex. **Yes,** we understand that we must change, **but** we have this and that problem and cannot.

When we conduct experiments to understand our systems' boundaries and reactions, we can improve our applications before an actual incident occurs. Chaos testing reduces incidents and stabilizes our systems. At the same time, we learn how our systems behave.

These are all measurable and tangible outcomes. We can measure the number of incidents, the incidents related to changes, and the time it takes to detect and solve an incident.

But SRE has more. With it, we are finally producing a system that can increase the number of customers it supports independently of the number of people who support the application. SRE improves customer satisfaction, which in turn increases revenue. For most IT departments, this is more challenging to measure. It requires close cooperation with the product owner and business stakeholders.

Our teams also see reduced fluctuations. Fewer people are leaving the company. Our approach to psychological safety in teams improves the work climate. Toxic behavior and leadership get called out. This makes cultural improvements tangible. Workplace satisfaction is an element that more and more companies want to improve, especially now that most companies are realizing that their employees have a voice. They find each other on social media and speak openly about the work culture, leadership behavior, and compensation.

To summarize, SRE drives the strategy that Agile and DevOps started. By increasing deployment frequency, it supports fast, small product iterations, improves quick rollouts, and reduces risk. SRE gives all team members the confidence to deploy quickly. It lets us push down the accelerator pedal and speed down the highway with the confidence that safety systems and high guardrails secure us.

Companies are coming to us and asking what their return on investment in SRE will be. What will they get back when they upskill their workforce, buy new tools, and give the time to work on optimizations for resiliency?

Some teams are starting to count incident tickets and calculate ad-hoc figures in a spreadsheet to show how much productivity they can save. Leadership is asking us what they should implement to reduce the number of incidents. We know that cost pressure is high.

We discussed this with our teams, interviewed them, and wrote down over 30 success stories about how they have saved time, reduced incidents, and reduced solution time. If we look at these cases and present them to other teams, the answer is always the same. These cases are always exceptional. They apply to one team because they had a unique situation, legacy tooling, noise in alerting, missing observability, or something else.

All the stories from the teams can be clustered under the SRE capabilities we listed above. However, each capability might not lead to the same outcome when we implement it. Our approach is to let the teams decide what makes the most sense for them. It depends on their situation, on their technological depth, on their skills, on their business requirements, and on customer needs. We see the best impact when we let teams decide what to do and don't force a spreadsheet from the top, requiring a number of ticket reductions.

Summary

Over the years, we've realized that in traditional companies, SRE is the solution to the problem you get when you deploy fast to production. It's a maturity step. First, companies must reach an understanding of the changes Agile and DevOps bring.

Agile changes the workflow to iterate faster. DevOps lets us automate deployments and testing. We do this because fast deployments reduce risk and allow us to optimize our product faster. Failing is part of the process. Nothing is optimal; we iterate, observe, learn, and act.

The message of SRE has stuck more with teams and companies that have already reached a higher deployment speed than quarterly deployments. The public cloud and the Cloud-Native movement are also helpful. However, many companies, especially in regulated industries, have just begun their journey. The result is a hybrid cloud, in which on-prem will always be part of a private cloud.

Our experience is also that teams welcome SRE. Many people in our industry want to change. They welcome new ideas. Primarily, when people have worked for a long time in operations, they see SRE as a new purpose. Operations people have felt left out in the past few years. Agile focuses on business and IT connections. DevOps lets developers, quality teams, and security teams improve. The cloud targets infrastructure teams. For years, we had no improvements in operations. Product teams welcome SRE; it puts their work in focus. SRE changes the dynamic and the organization in IT; it's the bridge between development, operations, and infrastructure.

We learned that the SRE concept is most effective when we combine build-run teams. SRE balances new features and stability. Without a shared understanding of the priorities between the teams and the option to pause new features to improve stability first, we lose a central part of the concept. This balancing is faster and easier when we share responsibility and work in a shared team.

One last learning point in this section is the risk classification of applications. The changes Agile, DevOps, and SRE bring let us rely heavily on tools. So far, business-critical applications have been primarily external customer-facing. An internal tool, for example, for deployments, was never essential for a company. But what happens if the DevOps pipeline is not working anymore? Most teams must stop deploying to production. Important bug fixes cannot be implemented. If the risk classification of a ticket is too low, it does not get the attention it deserves. This results in a longer solution time, which puts all of us at risk. It's an SRE's duty to understand these dynamics and to find solutions. In this dynamic and changing IT world, we must validate and question existing structures and established concepts, including risk classification.

Case Study

How do we reach our conclusions? Through experience. Our teams serve numerous clients, and with each iteration, we compare our understanding and optimize our transformation approach. Let's demonstrate this by looking at an example.

A high-tech company rolled out a new application from a vendor. The application shows the promotions, bonuses, and compensations for all the company employees. Its primary use is on a single day of the year, the compensation day. On this day, all the people in the company are getting informed about their bonuses. As you can imagine, expectations are high. Some people will decide their future on this day. For some, the future will be decided by the company. Some teams celebrate the compensation day together, all in the same room, with champagne and food. Later, the celebration moves on to appropriate locations where the boundaries with the company are not seen and felt anymore. This single day can stand for a whole year.

But this year was different. In the morning, everything was going fine. Management opened the application to inform one team member after another about their bonuses. The application crashed from time to time, but after a restart, everything was OK. Later in the day, everything was not okay. When the US colleagues started their day, the application crashed. It got restarted and crashed again and again and again. On this one day, everyone's opinion about internal IT was revised. This single story formed views for a very long time. Each time an application crashed, Outlook did not start, or a screen froze, it was water on the mills of their internal IT customers.

Why is this experience so fundamental for a book about SRE? Multiple aspects are worth mentioning:

- **Internal-Facing Applications Have Customers Too:** When validating the effort we want to put into resiliency, we must take our reputation and our employees into account. People move on when they do not get the right tools to do their job. For example, I'm using a dashboard to see my relevant metrics. These dashboards should give me the numbers about team and application performances in various slices. Most of these dashboards have the same issues; when I press a button, I have to wait. Sometimes, I wait so long that I forget what information I was searching for.

- **Risk Classification Is Dynamic:** Once a year, an internal application can be core to your business. Your business is always your internal reputation. The same can happen with an internal application that is only active once per quarter. Then, when your company closes the financials or at the end of the year, you do the annual accounting.

- **SRE Capabilities Must Be Considered in a General IT Operating Model:** Focusing only on toil, SLI/SLO, observability, and so on will only help you up to a certain point. We must consider all the elements contributing to the perception of stability and reliability. When an internal application crashes at a crucial moment, it sets the tone. Employers will transfer this experience to all applications. This makes our work challenging because we have to overcome these expectations.

Is SRE the solution to the problem we described above? SRE straight out of the book is not, and SRE for a startup is definitely not. But Enterprise SRE 2.0 is. Enterprise SRE 2.0 takes into consideration the operating model of the environment in which SRE lives. SRE is not only a specific capability but also an environment in which it is executed.

Enterprise SRE 2.0 has its SRE cosmos with Agile, DevOps, and the Cloud. Its legacy includes transitional thinking and traditional technologies. Its legacy also includes all the regulations experienced and written down in policies and IT processes. Employees interpret and implement policies in their daily struggles. This is the reality: Enterprise SRE 2.0 is reliably improving.

The key point for us is that we have to take resilience and stability seriously. These examples may be tiny, but crashes are setting the tone. An application crashing on the very day it should work becomes a story; a story we will hear again and again next to the coffee machines over and over again in the upcoming years. Taking only the financial considerations into account when we talk about SRE denies that the problem is more significant, and that, in some cases, the economic impact is not the main driver.

## CHAPTER 2

New Capabilities for the Tech Operating Model

In this chapter, we embark on a crucial exploration into the very essence of site reliability engineering (SRE) within the complex, interconnected world of the modern enterprise. We will not merely describe SRE; we will dissect its fundamental building blocks and illuminate the profound shifts it demands from the traditional tech operating model. More importantly, we will confront the significant hurdles organizations must clear to successfully integrate SRE, transforming it from a mere technical practice into a cornerstone of strategic business advantage.

The journey we are undertaking is not just about adopting new tools or processes; it's about a fundamental rewiring of how technology powers and sustains the entire business. It's about achieving innovation not despite reliability, but through it. This chapter will serve as your comprehensive guide, offering both the strategic insights necessary for executive understanding and the practical depth required for successful implementation, ultimately paving the way for technology to truly take center stage in driving business outcomes and continuous improvement.

The SRE Tech Operating Model

The year 2025 presents a unique and compelling confluence of forces—a "perfect storm"—that necessitates the rapid adoption of SRE principles within organizations. The digital landscape is shifting dramatically, driven by an ever-increasing velocity of customer expectations. Modern consumers, accustomed to the instantaneous and seamless experiences offered by digital native companies like fintechs and major consumer platforms, now demand nothing less from every service they interact with. This isn't merely a preference; it's a baseline expectation, a digital watermark against

F. Hoeppner and F. Sbaraglia, *Mastering Site Reliability Engineering in Enterprise*, https://doi.org/10.1007/979-8-8688-1448-8_2

which all services are now implicitly measured. For a legacy enterprise, failing to meet this benchmark means falling behind, losing market share, and eroding hard-won customer loyalty.

Simultaneously, the regulatory environment is tightening its grip, expanding both in scope and complexity. In the financial services sector, for instance, a labyrinth of data privacy laws, stringent operational resilience mandates such as DORA (Digital Operational Resilience Act), and critical anti-money laundering (AML) and know-your-customer (KYC) obligations now demand not just compliance but **unequivocal, demonstrable proof of system resilience**. This is no longer a tick-box exercise; it requires a deep, architectural understanding of system behavior under duress and the ability to proactively mitigate risks. Failure to comply carries not only hefty fines but also significant reputational damage, eroding trust with regulators and customers alike.

Adding another layer of urgency, the threat landscape from cybersecurity continues to evolve at an alarming pace. Criminal enterprises and even state-sponsored actors are employing increasingly sophisticated tactics, frequently targeting the very heart of the economy—financial institutions and critical market infrastructure. In this high-stakes environment, a single vulnerability can cascade into a catastrophic breach, impacting millions and jeopardizing the stability of entire markets. Therefore, robust system resilience is not just a technical aspiration but a national security imperative.

Against this backdrop, technology teams find themselves caught in a challenging paradox. They are under immense pressure to rapidly adopt bleeding-edge tools and platforms—from advanced AI capabilities to ubiquitous cloud environments—while simultaneously grappling with the immense burden of managing aging systems and entrenched legacy processes. It's like being asked to upgrade a moving airplane mid-flight while simultaneously building a new one. In essence, businesses are compelled to evolve with unprecedented speed, yet they must do so with uncompromised safety. They must relentlessly innovate, but never at the expense of unwavering reliability. And all of this must be achieved under the perennial, unforgiving pressure to control costs and optimize resource utilization. This intricate dance between innovation, reliability, and cost efficiency forms the strategic imperative for the SRE Tech Operating Model.

Why Traditional Models Break Down

In many organizations, the foundational problem lies in a deeply ingrained structural legacy: the work of building software and operating it in production remains stubbornly split across disparate, often siloed teams. Developers meticulously build the code, acting as the architects and craftsmen of new features. Operators, on the other hand, are tasked with the often-heroic, reactive effort of fixing issues once they manifest in the live production environment. Architects are responsible for the grand design, while risk managers diligently monitor for potential threats.

This traditional division, while perhaps a pragmatic solution in an era of simpler software systems and release cycles measured in months rather than minutes, has become a critical impediment in today's hyper-connected, continuously evolving digital world. It's akin to a factory where the designers never speak to the mechanics on the shop floor, and the mechanics only receive blueprints after a breakdown. In the era of continuous delivery, where code changes flow ceaselessly, and cloud-native architectures demand a fluid, dynamic approach to infrastructure, this separation between "build" and "run" creates perilous blind spots.

The critical flaw in this model is the **fragmentation of ownership and visibility**. When an issue arises, it's not uncommon for the problem to become a "hot potato," tossed between teams. The development team might argue it's an operational issue, while operations might contend it's a fundamental design flaw. This leads to frustrating delays, finger-pointing, and a pervasive lack of clear accountability. Crucially, **no one individual or team possesses full end-to-end visibility or ownership of reliability**. The "last mile" of reliability, the actual user experience in production, becomes an orphaned responsibility, leading to slower incident response, reduced system resilience, and a palpable erosion of trust, both internally and with external customers.

SRE emerges as the powerful antidote to this fragmentation. It represents a fundamentally different paradigm, unifying engineering, operations, and even business teams under a single, overarching, shared goal: **the delivery of reliable, scalable systems that users can inherently trust**. This isn't merely about tweaking existing processes; it demands a wholesale transformation, embracing not only new tools and roles but, more profoundly, new capabilities and an entirely reimagined operating model. This new model is characterized by one paramount principle: **it places resilience firmly at the center of how software is conceptualized, planned, built, delivered, and run**. It shifts the focus from merely reacting to failure to proactively engineering for enduring reliability, making it an intrinsic feature, not an afterthought.

Building a Foundation of Reliability

Whether a capability like SRE is entirely novel to an organization or represents an evolution of existing practices, the broader spectrum of reliability-focused practices directly shapes and influences the Tech Operating Model. This model serves as the architectural blueprint for how technology effectively powers, supports, and differentiates business operations. It is not a static document but a living framework, encompassing five critical dimensions that we will explore in detail, each serving as a pillar supporting the edifice of enterprise reliability.

These dimensions provide a holistic view of the interconnected elements required to embed SRE successfully:

- **Organizational Structure and SRE Roles:** This dimension precisely defines the positioning of SRE within the broader technological landscape of an organization. It grapples with fundamental questions: Is SRE a standalone, dedicated central team, acting as an internal consultancy group, or are SREs directly embedded within development and/or product teams, functioning as a specialized role or a foundational skill set across all engineers? Beyond placement, it delineates the core responsibilities of SREs, which include, but are not limited to, defining Service Level Objectives (SLOs), systematically automating repetitive operational tasks (toil), and leading incident response efforts. It also emphasizes the critical interdependencies and collaboration models required, for example, how SREs work alongside development teams and product managers on crucial concepts like error budgets, and how they articulate the tangible impact of reliability to other business units. This structural clarity is paramount for effective SRE adoption.

- **Reliability-Driven Processes and Workflows:** This dimension meticulously outlines the standardized methodologies and systematic workflows designed to embed and ensure reliability across every phase of the software lifecycle. It encompasses SRE's pivotal involvement in production readiness reviews, the meticulous definition of robust incident management protocols, the implementation of blameless postmortem procedures, and the strategic integration of change management processes fortified with automated reliability gates. Furthermore, it highlights the continuous delivery practices that, by design, prioritize stability and robustness in lockstep with the imperative for speed, recognizing that true agility is built on a bedrock of reliable systems.

- **Reliability Technologies and Tooling:** This dimension specifies the foundational technology stack and the array of specialized tools that are indispensable for enabling and sustaining reliability. It underscores the absolute necessity of being equipped with the right technological infrastructure to support new capabilities, acknowledging that in most areas, successful SRE implementation heavily relies on these tools. This includes, but is not limited to, standardized observability platforms—encompassing comprehensive monitoring, logging, and distributed tracing—which provide consistent, real-time visibility into system behavior. It also covers robust automation frameworks for the reduction of toil and the creation of self-healing systems, sophisticated chaos engineering tools for proactive resilience testing, and the pervasive adoption of Infrastructure as Code (IaC) to ensure consistent, reliable, and reproducible infrastructure deployments.

- **SRE Capabilities and Culture:** This dimension moves beyond mere technical skills to focus on cultivating the deeper aptitudes and the critical mindset essential for achieving sustainable, long-term reliability. It emphasizes fostering a blameless learning culture, where system failures are not seen as opportunities for assigning blame but as invaluable opportunities for systemic improvement. It also addresses the imperative of developing robust software engineering skills within traditional operations teams, promoting a proactive, forward-looking approach to risk management, and instilling a continuous, ingrained drive to automate away repetitive tasks, often referred to as "toil." This cultural shift is arguably the most challenging, yet most impactful, aspect of SRE adoption.

- **Reliability Performance Metrics:** This dimension defines the critical framework for how SRE success and overall system reliability are quantitatively measured, tracked, and reported across the organization. It centers on the rigorous use of Service Level Indicators (SLIs) and Service Level Objectives (SLOs) to precisely define desired reliability targets, providing a clear, unambiguous benchmark for performance. It also encompasses the strategic management and utilization of Error Budgets, which offer a quantitative mechanism for balancing the relentless drive for new features with the non-negotiable demand for stability. Finally, it includes the reporting of key operational efficiency metrics such as Mean Time To Restore (MTTR) and the ratio of toil versus strategic engineering work, providing data-driven insights into operational health and SRE effectiveness.

When introducing a transformative new capability like SRE, the strategic objective is not to embark on a radical, wholesale rewrite of the entire Tech Operating Model. Such an approach would be disruptive, time-consuming, and often counterproductive. Instead, the goal is to **judiciously adapt the existing model**, leveraging its strengths while strategically augmenting it with SRE principles. To achieve this adaptation effectively, the crucial first step involves a comprehensive identification of precisely which areas of the existing model require adjustment, enhancement, or fundamental change.

Tech Operating Models, by their very nature, are equipped with a number of inherent properties that determine their effectiveness and strategic alignment. Firstly, they must be **aligned with the overarching business strategy**, ensuring that technology serves as a proactive enabler and accelerator of the organization's strategic objectives, rather than merely a cost center. Secondly, they are designed to drive **efficiency and effectiveness**, optimizing the utilization of technology resources to streamline business operations and improve critical outcomes. Thirdly, in today's volatile environment, they must embody **flexibility and agility**, allowing organizations to rapidly adapt to unforeseen technological shifts, evolving market dynamics, and emergent business requirements. In rapidly changing environments, the ability to continuously monitor risk and respond proactively is paramount for maintaining resilience. The tech operating model, therefore, must inherently support the identification and mitigation of potential technology-related dangers and challenges, transforming vulnerabilities into strengths.

A core outcome of a well-defined tech operating model, particularly when infused with SRE principles, is the facilitation of **scalability and reliability**. It provides the necessary framework for technology capabilities to grow and expand seamlessly with business demands. To ensure this, the model mandates rigorous **monitoring and measurement**, implementing a comprehensive suite of metrics to track and assess its performance. This commitment to data-driven decision-making empowers the technology department to engage in transparent, fact-based communication and collaboration with the business. This, in turn, significantly enhances coordination between technology and other business functions, elevating technology to a central, strategic position within the company, capable of driving not only its own continuous improvement but also directly influencing overall business outcomes. The aspiration is to establish a robust framework for ongoing refinement and innovation, incorporating insights derived from failures, successful experiments, shifts in customer behavior, evolving ecosystem providers, and continuous technological advancements, such as the transformative impact of Generative AI on resource workflow and allocation.

Ultimately, the primary outcome of the Tech Operating Model is to ensure the efficient utilization of technology resources for core business purposes. However, its stakeholders extend beyond just the internal business units. External stakeholders, including critical areas like compliance and security, are also a primary focus. The model must therefore incorporate measures that proactively meet regulatory requirements and rigorously ensure data security, recognizing that trust and integrity are non-negotiable.

What's fundamentally important to grasp is that SRE will not only reshape *what* we communicate to the business but, more profoundly, *how* we communicate it. The narrative around customer experiences with system reliability will become significantly more detailed, more frequent, and far more nuanced. As champions and enablers of SRE, we must acutely recognize this strategic shift and proactively take an active, guiding role in shaping this new communication paradigm. This transformative effort begins with a deep, candid understanding of our own organization's inherent tolerance for risk—a concept that underpins the entire approach to SRE adoption.

Risk Tolerance Assessment for Organizational Change

Understanding an organization's risk tolerance is not merely an academic exercise; it is the linchpin that dictates the entire approach to rolling out any new capability, particularly one as conceptually challenging as SRE. This is especially true for practices that may sound inherently risky, such as "chaos engineering" or the concept of "error budgets," which can fundamentally alter entrenched thinking about reliability. As a seasoned practitioner, I've witnessed firsthand companies prematurely abandoning promising SRE initiatives simply due to the perceived implications of a name or a misunderstanding of underlying principles.

For organizations with a **high-risk tolerance**, the path to SRE adoption can be accelerated. This agility allows for faster experimentation and bolder implementation strategies. Conversely, in **risk-averse companies**, a more deliberate, measured approach is necessary, requiring a greater investment in organizational change management. In extreme cases, one might even consider renaming concepts like "Chaos Engineering" or "Error Budget" to something more palatable and culturally aligned, such as "Stability Testing," "Risk Identification Strategy," or "Proactive Anomaly Discovery." While renaming should always be a last resort, as adhering to industry-standard naming conventions is generally preferable, the imperative to balance this with the company's unique cultural fabric is paramount for successful adoption. As a change management specialist, my primary objective is to meticulously evaluate my company's risk tolerance and, armed with that insight, strategically determine the specific activities and interventions required to ensure successful adoption, potentially even shortening the time-to-value for SRE initiatives.

As we embarked on planting the seeds of SRE thinking within diverse enterprises, a consistent insight emerged: **risk tolerance and the understanding of risk vary significantly, not only by industry but even from company to company within the same sector**. The underlying perspective on risk can be deeply rooted in several interconnected dimensions:

- **The Industry Itself:** The fundamental nature of an industry profoundly shapes its perception of risk. A global financial institution, for example, views risk through an entirely different lens than a niche coffee roaster. An insurance company inherently calculates and manages risk as its core business, while an entity like NASA operates as an undisputed expert in mitigating catastrophic risk. Certain industries, by virtue of their rewards, regulatory scrutiny, and inherent obligations, naturally cultivate a higher or lower baseline risk tolerance. For instance, a fast-paced e-commerce platform might tolerate more volatility than a critical healthcare system, simply due to the differing consequences of failure.

- **Maturity and Stage of Growth:** An organization's lifecycle stage plays a crucial role. A nascent startup operating in the financial services sector might be inclined to embrace more risk than a venerable, century-old global bank. The early-stage company, typically with less established infrastructure and fewer legacy obligations to shareholders, might be incentivized to move faster and embrace higher risk to rapidly establish market presence and achieve accelerated growth. Shareholder expectations for rapid innovation and disruption often align with a higher risk appetite in these scenarios. Conversely, the established 100-year-old bank, having navigated countless economic cycles and potentially incurred significant penalties for past missteps, tends to possess a profoundly low-risk appetite. Its entrenched structure and long history often compel it towards increasingly conservative postures, prioritizing the protection of its vast, accumulated assets and established reputation.

- **Capital Reserves:** The availability of capital acts as a direct enabler or constraint on an organization's risk posture. A startup, particularly one operating in a high-growth sector like Generative AI in late 2023, might enjoy exceptional access to funding, providing them with ample spare capital. This financial buffer allows them to absorb potential failures and recover from setbacks, thereby affording a higher tolerance for risk. Conversely, organizations without substantial cash reserves or robust funding sources are naturally more constrained and less able to absorb the financial consequences of risk-taking.

- **Leadership Preference:** The individual tendencies and strategic leanings of the company's leadership—its managers and executives—exert a powerful influence on its overall risk preferences. A new leader, eager to establish their presence and drive change, might actively advocate for and champion new ideas, potentially promoting more aggressive practices like Chaos Engineering. Conversely, a leader nearing retirement, with an established reputation built on stability and predictability, might be more inclined to resist

significant change and maintain the status quo. Their personal
comfort with uncertainty directly translates into the organization's
strategic appetite for risk.

- **Company Culture:** Beyond individual leaders, the deeply ingrained
  company culture shapes its collective attitude towards innovation
  and risk. Some organizations inherently foster an innovative,
  entrepreneurial culture—Xerox, for instance, is renowned for its
  historical drive to innovate. Others, by design or evolution, become
  bureaucratic and inherently conservative, prioritizing process and
  control over rapid experimentation. This cultural disposition forms
  the bedrock upon which all risk decisions are made.

To strategically align SRE practices with an organization's unique profile and
inherent risk DNA, we have developed a concise yet potent tool: the **Short-Risk
Tolerance Assessment**. This framework provides a structured, analytical pathway
to explore where a company positions itself across five critical dimensions. Each of
these dimensions—**leadership, innovation culture, regulation, market dynamics,
and financial reserves**—directly correlates with and influences how an organization
formulates reliability decisions, how it fundamentally responds to system failures, and
how it strategically balances the twin imperatives of stability and speed. By gaining a
profound understanding of your organization's standing across these axes, you can
precisely tailor your SRE implementation strategy to seamlessly fit within—rather than
aggressively fight against—your organization's core DNA, thereby maximizing adoption
and impact.

Here is the **Short-Risk Tolerance Assessment** framework:

	High Risk Tolerance	Medium Risk Tolerance	Low-Risk Tolerance	Key Question
Leadership	We want to have breakthrough innovations despite the uncertainty.	We strive for balance with our core business when we pursue emerging opportunities.	Our resources focus on improving current products/services and making small iterations.	How comfortable is the leadership team with making decisions that have a high degree of uncertainty in outcomes?
Innovation Culture	We have a constant pipeline with innovation, a team testing new capabilities with pilots to test new ideas and concepts.	Occasionally, we test new ideas and capabilities while optimizing current models.	Carefully analyses ROI before allocating resources to new ideas, a business case drives decisions.	How fast is your company able to prototype and test new product/service concepts?
Regulation/Compliance	We challenge legal grey areas despite pushback to be at the forefront.	Adhering closely to regulations intent and clarity, we conduct regular internal audits.	We go beyond mandatory compliance requirements to give our customers a highly secure experience, and we set safety standards for our industry.	How strictly does your company adhere to regulations versus aiming at the edge of what is permissible?
Market Dynamics	We shape and disrupt our industry with bold investments.	We respond nimbly to shifts in the business landscape.	We carefully protect our current market position.	How disruptive are emerging technologies and shifts in your business environment?
Financial Reserves	We invest most of the available capital to fuel our growth.	Our company maintains a moderate buffer as needed.	Conserves substantial reserves to ensure stability.	How many months could your company sustain operations if a major disruption halted incoming revenue?

Figure 2-1. *Short-Risk Tolerance Assessment Framework*

Based on the insightful results derived from this risk tolerance assessment, I strongly recommend a tailored approach to adjusting your change management activities when implementing SRE within your organization. This strategic customization is vital for securing necessary buy-in, accelerating adoption, and ultimately maximizing the positive outcomes from your SRE initiative. Specifically, you will want to consider:

- The **level of sponsorship from leadership** you may require. In high-risk tolerance environments, a lighter touch might suffice, whereas low-risk tolerance settings demand visible, active, and consistent executive championship.

- Whether to pursue **small, incremental steps or embrace the full scope of SRE** from day one. This decision is directly correlated with the organization's comfort level with change and disruption.

- Your chosen approach to **communication**, specifically whether to be fully transparent and engage in extensive over-communication. In low-risk tolerance organizations, clear, frequent, and reassuring communication is a non-negotiable imperative.

Recommended Activities Based on Risk Tolerance

High-Risk Tolerance Organization: For organizations that embrace high levels of risk, your SRE implementation strategy can be bold and fast-paced, aligning with their inherent drive for innovation and rapid growth:

- **Position SRE As a Direct Accelerator for Business Value:** Frame it as a powerful enabler that boosts the deployment of new features into production even faster, directly linking reliability to speed of innovation.

- **Incentivize Early Adopters and Pilot Teams:** Provide tangible incentives to teams that willingly opt in and actively participate in SRE pilot programs, fostering enthusiasm and a sense of shared success.

- **Showcase Examples of Advanced Innovation Companies:** Communicate compelling case studies of other leading-edge organizations that have successfully leveraged SRE to achieve significant breakthroughs, leveraging external validation.

- **Measure and Reward System Resilience:** Implement clear metrics for system resilience and establish reward mechanisms for teams that consistently deliver highly resilient systems, integrating reliability into performance recognition.

- **Rapidly Scale Successful Pilots to Production:** Empower teams with self-service capabilities from day one, enabling a rapid transition of successful SRE pilots into full production adoption.

- **Embrace Learnings from Failures with Agility:** Foster a culture where failures are viewed as invaluable learning opportunities, driving rapid iteration and continuous improvement.

Moderate-Risk Tolerance Organization: For organizations with a balanced approach to risk, the strategy involves a careful blend of cautious progress and clear communication, building trust and demonstrating value incrementally:

- **Acknowledge and Address Hesitations:** Proactively acknowledge any inherent hesitance towards unnecessary risks or unstructured experimentation, validating concerns and building confidence.

- **Clearly Differentiate Environments:** Communicate the distinct differences between developer, quality assurance, and production environments, outlining the plan for their strategic use in SRE implementation.

- **Conduct Early Town Halls on Due Diligence and Safety:** Organize forums to transparently discuss the rigorous due diligence processes and safety precautions embedded within SRE practices, particularly for concepts like chaos engineering.

- **Start with Small, Low-Impact Applications:** Begin SRE adoption with less critical applications, focusing on learning and demonstrating value in a controlled, contained environment.

- **Outline Clear Escalation Procedures:** Establish explicit procedures for escalation, including named individuals and their responsibilities, and create structured reports on findings, with clear criteria for halting experiments if necessary.

- **Progress Incrementally with Transparency:** Move from pilots to targeted deployments in small, highly transparent increments, ensuring clear, consistent communication at every step.

Low-Risk Tolerance Organization: In organizations characterized by a low appetite for risk, the SRE adoption strategy must emphasize safety, control, and a strong focus on risk identification and mitigation:

- **Position SRE As an Early Technical Risk Identification Strategy:** Frame SRE as a crucial tool for proactively identifying technical risks and evaluating service consistency, aligning it with their core conservative values.

- **Communicate Early, Often, and with Absolute Clarity:** Engage stakeholders frequently and transparently, ensuring all communications are unambiguous and reassuring, focusing on the safety and control aspects.

- **Lead Communication with Success Stories, but Also Provide Transparent Insights on Challenges:** Highlight successful SRE implementations internally and externally, but also include sections detailing challenges encountered and the robust mitigation strategies employed.

- **Customize Simulations to Address Specific Pain Points:** Tailor chaos engineering simulations to directly address well-known vulnerabilities or historical major outages within the company, demonstrating direct relevance and value.

- **Implement SRE Techniques Progressively, One Application at a Time:** Adopt a highly iterative, cautious approach, requiring formal approval before proceeding to the next application or phase of implementation.

- **Define Strict Criteria and Oversight for Experiment Approval:**
  Establish rigorous approval processes and provide strong oversight
  for all SRE-related experiments, ensuring maximum control and
  minimizing perceived risk.

- **Provide an Opt-Out Opportunity and Feedback Channels:** Offer
  teams the option to opt out of early SRE initiatives and create clear
  channels for submitting feedback, fostering a sense of control and
  collaboration.

- **Report Audit Trails of Controlled Experiments and Their Impact:**
  Maintain meticulous audit trails of all controlled experiments,
  documenting their impact and outcomes to demonstrate
  accountability and control.

Organizational Structure and Roles

Let's reimagine our technology organization as a dynamic ecosystem where reliability
is not an afterthought but rather deeply woven into the very fabric of every stage of the
software lifecycle. It is no longer something that teams frantically scramble to fix *after*
an incident has disrupted services; instead, it is an intrinsic consideration that begins
on day one, the very moment we commence designing and building new software. This
proactive approach is foundational to the SRE paradigm.

- In the **planning and design phase**, teams proactively initiate the
  process by meticulously identifying the precise level of reliability
  their systems require. This critical step involves setting clear,
  measurable Service Level Objectives (SLOs), carefully estimating
  potential risks, and gaining a profound understanding of what failure
  might look like from the nuanced perspective of the end-user. By
  laser-focusing on critical user journeys—the pathways most vital to
  the customer experience and business value—teams can strategically
  prioritize what genuinely matters and design for resilience in those
  high-impact scenarios. This shifts the mindset from generic uptime
  targets to user-centric reliability.

- During **development**, engineers intentionally employ patterns and practices specifically designed to render systems more inherently resilient. This includes writing comprehensive tests that deliberately simulate various failure modes, judiciously using feature flags to enable gradual and controlled rollout of new functionality, and architecting systems with an inherent capability to tolerate the loss of individual nodes or even entire regions. The fundamental shift here is profound: instead of merely *hoping* that systems will not break, the SRE mindset **assumes they *will* break—and meticulously prepares for it**, building in layers of redundancy and recovery mechanisms from the outset.

- As code progresses towards **deployment**, teams leverage sophisticated techniques such as canary releases or blue/green deployments to systematically minimize risk. Monitoring hooks are meticulously embedded to track system behavior in real-time, providing immediate feedback. Crucially, if something goes awry, automated rollbacks are configured to rapidly revert systems to a known safe state, preventing widespread impact. This is about building intelligent gates, not just manual checkpoints.

- Once systems are live in **production**, observability transcends mere uptime checks; it becomes absolutely essential. Teams require comprehensive visibility not just into whether systems are technically "up," but whether they are genuinely *working as users expect*. Alerting mechanisms are intelligently configured based on meaningful SLOs, rather than arbitrary technical thresholds, ensuring that alerts are actionable and tied to business impact. When incidents inevitably occur, on-call engineers respond with agility, and critically, the team conducts a blameless retrospective afterward, transforming each failure into a collective learning opportunity to understand the root cause and systematically improve the system.

- And over time, the journey of optimization continues relentlessly. Teams continuously strive to reduce repetitive manual work, often termed "toil," through strategic automation. They continuously enhance automation capabilities and leverage insights gleaned from past failures to make systems progressively stronger and more robust. This iterative refinement embodies the principle of continuous improvement, turning every incident into a step forward.

We now understand that the burden of reliability is not confined to a single team or individual; it is a shared responsibility, diffused across numerous teams throughout our technology organization. Indeed, the imperative of tech resiliency extends its weight across many shoulders. For too long, this responsibility has been fragmented, and in such a scenario, the classic adage often holds true: when it's everyone's burden, in reality, no one truly owns it. This decentralized approach, while seemingly adequate for many companies in the past, has reached its breaking point with the increasing complexity of modern systems, leading to pervasive failures.

In today's highly complex and dynamic tech landscape, a technology organization critically needs a central figure or function—an "owner on top of the house"—to effectively manage and continuously improve overall resiliency. In a **legacy company tech organization**, as depicted in the provided graphic, the Chief Information Officer (CIO) typically has direct reporting lines from architects, application engineering, infrastructure, operation teams, and so forth. In this structure, the burden of resiliency is often scattered across these disparate teams, leading to fragmented responsibility. Cybersecurity teams are primarily concerned with external threats, while developers prioritize shipping new features at speed. Business teams, understandably, expect everything to "just work." Each group performs its best to manage its particular piece of the puzzle, but in the absence of a single, unifying owner, the collective effort can feel disjointed, inefficient, and often reactive.

While one might argue that the CIO ultimately "owns it all," in the pragmatic reality of a large enterprise, the strategic burden of reliability must be managed proactively, a step ahead of reactive CIO oversight. This is achieved through a clearly defined, cross-functional reliability governance model, robustly anchored by site reliability engineering principles. This signifies a profound shift: proactively embedding reliability accountability directly within product teams, who are then powerfully supported by specialized SRE coaches, clear Service Level Objectives (SLOs), and shared error budgets. Instead of the CIO primarily reacting to outages after they have already occurred, the organization cultivates a pervasive culture where reliability is treated as a first-class feature—an intrinsic quality jointly owned by both business and technology leaders. This model incorporates sophisticated mechanisms to surface, prioritize, and systematically remediate risks *before* they escalate into full-blown disruptions.

In this optimized, future-ready model, the CIO gracefully transitions from being the sole gatekeeper of reliability to becoming a dynamic enabler of a system where resilience is truly everyone's responsibility—a responsibility ingrained structurally, culturally, and operationally.

To holistically address this fragmentation, the company's organizational structure must strategically evolve. A more modern, highly resilient setup would centralize accountability for technology resilience under a single, dedicated individual, ideally reporting directly to the CIO. This leader would be vested with a clear, overarching mandate to oversee technology resilience from end-to-end, ensuring faster, more informed decision-making, superior proactive preparation, and demonstrably more effective responses to disruptions. Concurrently, this individual would maintain close, albeit informal ("dotted line"), collaborative connections to the Chief Risk Officer and the Head of Business Operations. These crucial relationships ensure that while the resilience leader holds primary responsibility, robust collaboration across key risk and business operations functions remains active and synergistic, preventing siloing and fostering a unified approach to organizational resilience.

The comparative graphics of "Legacy Company Structure" versus "Elevated Resiliency Organization Structure" vividly illustrate this transformative shift. In the latter, a dedicated "Resiliency" function, typically led by a Chief Reliability Officer or a similar role, is positioned as a peer to other critical functions like architects, application development & engineering, infrastructure & operations, data & analytics, and cybersecurity. This elevates reliability to a strategic domain, giving it the necessary executive visibility and mandate to drive change across the enterprise.

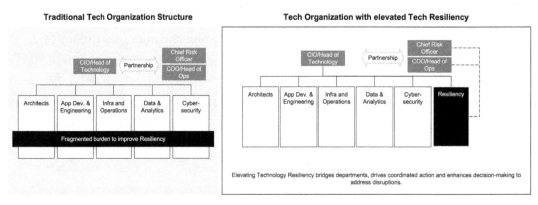

Figure 2-2. Tech Organization Structure

There are unequivocally clear strategic advantages to this elevated model. Concentrating responsibility for resilience under a single, accountable role creates profound clarity, significantly accelerates reaction times during incidents, and systematically fosters a pervasive culture of proactive risk management throughout the organization. Moreover, it provides executives with a singular, authoritative point of contact for all inquiries related to resilience, eliminating the need to navigate a bewildering maze of disconnected teams. This streamlined accountability translates directly into more efficient governance and clearer strategic direction.

However, it is crucial to acknowledge that this progressive approach is not without its inherent challenges. Centralizing authority, if not managed meticulously, can occasionally lead to bottlenecks, particularly if the responsible individual becomes overwhelmed or lacks adequate support and empowerment from development and operations leads. It also demands careful and continuous attention to ensure that the vital collaboration with risk and operations teams remains actively engaged and does not, through oversight, become a mere afterthought. The risk is that a new silo is created if the dotted lines become truly invisible.

Despite these potential challenges, the strategic benefits derived from elevating resilience to a focused, highly accountable role almost invariably outweigh the inherent risks. In today's relentlessly interconnected world—where technology disruptions can directly and instantaneously impact a company's hard-earned reputation and its financial bottom line—it is no longer sufficient to merely spread responsibility thinly and diffusely across the organization. Building a clear, strong, and dedicated backbone for resilience is far more than a mere technical upgrade; it is a fundamental strategic imperative.

When we successfully bring reliability under a framework of shared, explicit responsibility, we forge an organization truly capable of striking the delicate yet crucial balance between speed and stability. This mature setup enables a seamless, intelligent shifting of focus and capacity between development imperatives and technology operations demands as required. Such an organization moves with impressive velocity because it operates with an inherent, unwavering trust in the fundamental solidity of its production systems—a trust that empowers it to take calculated, smart risks with the rollout of new features and rapid deployments. And critically, when circumstances demand, it possesses the agility to instantly hit pause, reallocate resources, and refocus its collective energy on systematically strengthening the system's underlying resilience. This adaptability is the hallmark of a truly resilient enterprise.

Resiliency As a Top-Level Domain

One of the most frequent and critical questions posed by leadership when embarking on an SRE journey concerns its precise placement within the organizational hierarchy. In most traditional companies, structures remain deeply entrenched along the lines of distinct development and operations departments. The fundamental dilemma then arises: Does this new capability—SRE—need to be aligned with one side or the other, or should it be placed under infrastructure, perhaps as a specialized team? As we have just established, the strategic imperative dictates that reliability, and by extension SRE, must logically "sit on top of the house." This top-level positioning, particularly when a company is initiating SRE from the ground up, is absolutely key to its success and influence.

Once this strategic placement is secured, the next logical question arises: What precisely falls under this new mandate of reliability, and how should it be meticulously structured to deliver its full value?

Under this overarching "Resiliency as Top-Level Domain," we have meticulously defined a hierarchical structure of capabilities: Level 1, Level 2, and Level 3. This layered approach ensures strategic alignment, manageable focus areas, and clear accountability at the operational level:

- **L1 (Level 1)** capabilities represent the highest-level strategic pillars, serving as overarching frameworks to align executive priorities and investments. An example would be "Foundational Elements for Reliability" or "Proactive and Preventative Practices (Avoiding Failures)." These are the broad strategic categories that resonate with C-suite objectives.

- **L2 (Level 2)** capabilities group related practices into more manageable focus areas. These are typically designed for portfolio roadmaps, strategic funding allocation, and clear ownership assignments within the organization. For instance, under "Foundational Elements," you might find L2 categories like "Toil Reduction" or "Infrastructure & Platform Reliability." These provide a logical grouping for related workstreams.

- **L3 (Level 3)** capabilities detail the actual, fine-grained practices and specific activities that engineers and operations teams perform on a daily basis. These are mapped directly to specific teams and roles for granular accountability, ensuring that theory translates into practical action. Under "Toil Reduction," for example, an L3 capability would be "Toil Management." These are the tangible actions and technical tasks.

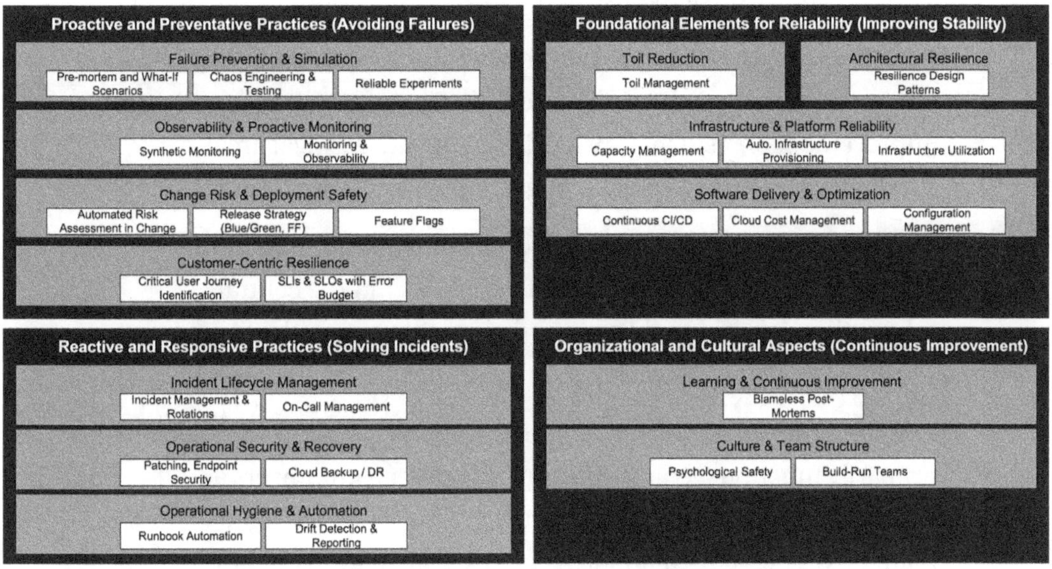

Figure 2-3. *Level 1–Level 3 Reliability Capability Map*

The "Level 1–Level 3 Reliability Capability Map" visually represents this hierarchical breakdown, organizing capabilities into key thematic areas such as "Proactive and Preventative Practices (Avoiding Failures)," "Foundational Elements for Reliability (Improving Stability)," "Reactive and Responsive Practices (Solving Incidents)," and "Organizational and Cultural Aspects (Continuous Improvement)." This structured categorization provides a comprehensive overview of all the building blocks necessary for a robust SRE implementation.

The pivotal decision point for bringing these capabilities to life—initially outlined in Chapter 1—revolves around the three primary options for placing the SRE capability itself. This choice profoundly influences how the resiliency as a top-level domain operates and, crucially, the required team capacity within it:

- **SRE As a Full-Time Role in a Squad:** In this model, the top-level Resiliency Domain acts as a provider of dedicated SRE practitioners, who are then deployed long-term and embedded directly into specific product and platform squads.

  - **Implication for Resiliency Top-Level Domain:** This model requires a **high** number of people within the central Resiliency Top-Level Domain, as they are essentially a talent pool supplying specialized SRE expertise across the organization.

- **In a Central Hub:** Here, the Resilience Top-Level Domain deploys SRE practitioners temporarily to squads. Their focus is on operational improvement of resiliency, prioritizing the most critical areas, conducting reviews, and often engaging in hands-on improvements.

  - **Implication for Resiliency Top-Level Domain:** This model requires a **medium** number of people within the central Resiliency Top-Level Domain, as practitioners cycle through different teams, sharing expertise more broadly.

- **As a Skill for All:** In this model, the Resiliency Top-Level Domain primarily functions as an educator and guide. It empowers and trains other practitioners who reside outside its direct organizational boundary. Their core mandate is to drive the maturity of SRE capabilities across the entire firm, systematically reducing bottlenecks and championing innovation topics.

  - **Implication for Resiliency Top-Level Domain:** This model requires a **low** number of people within the central Resiliency Top-Level Domain, as its focus shifts from direct operational engagement to enablement and thought leadership.

Irrespective of the chosen operational model, a universal truth for SRE success is that **SRE teams must be deeply integrated and collaborate seamlessly with other teams**, both within and outside the company. This imperative extends to external entities, recognizing the value of understanding and sharing best practices from industry peers. Inter-team cooperation is not merely desirable; it is a critical accelerator, enabling teams to work faster and more securely. To foster this essential collaboration, organizations must actively provide the right tools, establish meaningful incentives, and eloquently articulate the intrinsic value of inter-team cooperation. This includes establishing a robust platform for sharing knowledge and insights across the entire organization. Furthermore, organizing a vibrant **community of practice** is crucial to facilitate the sharing of learnings and insights from experiments and to continually motivate teams, thereby building a collective intelligence around reliability.

The "Resiliency Domain structure based on capabilities" graphic illustrates a sophisticated organizational model for the Resiliency Top-Level Domain team. In this structure, the Level 2 domains (e.g., "Proactive and Preventative," "Foundational Elements for Reliability") function as distinct streams. They are responsible for defining standards within their respective capability areas and effectively act as product owners for those capabilities, ensuring consistent implementation and evolution.

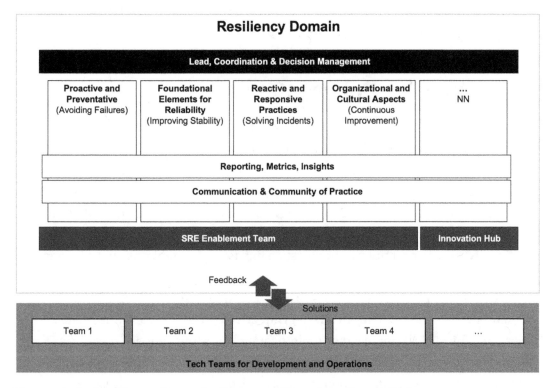

Figure 2-4. *Resiliency Domain Structure Based on Capabilities*

A key operational component is the **SRE Enablement Team**. This team serves as the crucial interface, interacting directly with other core technology teams (the "squads") and providing essential support to developers and operations specialists from a resilience perspective. The size and composition of this enablement team will vary significantly based on how the resiliency domain is operationalized—whether SRE is a full-time role embedded in squads, managed from a central hub, or championed as a skill for all.

Complementing this, the **Innovation Hub** plays a vital external-facing role. This team is tasked with continuously bringing in new capabilities, cutting-edge technologies, and innovative tools from outside the company. They run controlled pilots, and if successful, work to make these innovations accessible and actionable for all relevant teams across the organization.

Finally, the **Reporting, Metrics, and Insights** layer, along with the **Communication & Community of Practice** layer, cut across all these functions. They are responsible for building and nurturing the community, discovering actionable insights from data, and undertaking critical leadership and team reporting, ensuring transparency and data-driven decision-making throughout the entire resiliency domain.

SRE As the Glue Between Teams

Having meticulously described the internal organization of the Resiliency Domain, it is crucial to understand how its core component, the **SRE enablement team**, acts as a vital conduit, providing solutions and expertise to the various "squads" within the broader tech organization. These "squads" represent agile, cross-functional teams, each typically responsible for developing and maintaining specific technical products or platforms. The SRE enablement team plays a pivotal role in unifying the organizational perspective on reliability, reframing it not as a binary state but as a dynamic, continuous balance between the imperatives of speed and stability.

This unification is achieved by establishing a shared framework for decision-making. For example, the concept of **Service Level Objectives (SLOs)** elegantly translates intricate technical performance metrics into tangible business impact, creating a common language that bridges the gap between engineering and business stakeholders. Furthermore, **Error Budgets** introduce a powerful, quantifiable mechanism for balancing innovation with risk. This means if a service is consistently meeting its predefined reliability goals, the responsible team has the strategic latitude to take on more calculated risks, perhaps accelerating new feature development or experimenting with novel technologies. Conversely, if the service is falling short of its reliability targets, the team is empowered—and indeed, obligated—to pause feature work and proactively invest in stability improvements. These foundational concepts provide everyone, from front-line engineers to executive leadership, a common, objective language for discussing and making informed trade-offs, ensuring that reliability is always a conscious, data-driven choice, rather than a reactive consequence. We will delve into these concepts in much greater detail later in the chapter.

Bringing all these elements together within the holistic tech operating model, we identify crucial inflection points where the core tenets of resilience, stability, and robustness in technology are meticulously assembled and continuously reinforced. These are primarily the agile "squads" working in concert with the refined processes, where reliability is not merely a goal but an active, continuous act of creation.

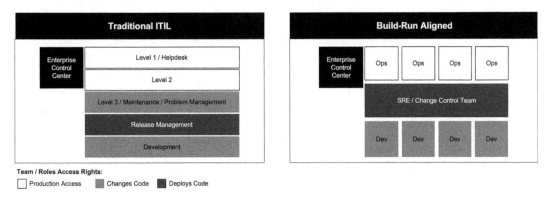

Figure 2-5. *Squad Structure Example, Traditional ITIL Team Structure Versus Build-Run Team Structure*

These squads can be fundamentally structured in two distinct ways, each with its own operational philosophy: the **traditional ITIL-based model** or the more modern **build-run team approach**. The "Squad Structure: Traditional ITIL team structure versus Build-Run team structure" graphic vividly illustrates these contrasting organizational paradigms. The traditional ITIL model often depicts a hierarchical structure with distinct layers like Enterprise Control Center, Level 1/Helpdesk, Level 2, Level 3/Maintenance/ Problem Management, Release Management, and Development, often leading to handovers and silos. In contrast, the Build-Run Aligned model shows a flatter structure where Development and Operations are integrated within squads, with an SRE/Change Control Team acting as a central enabler.

As we can observe from the following tables, a diverse array of teams are involved in and directly impacted by the strategic rollout of SRE capabilities, highlighting the cross-functional nature of this transformation.

Proactive Reliability Measures (Prevention)

These capabilities are the essence of foresight in SRE, focusing on anticipating and preventing potential issues before they can escalate and impact end-users. They embody a forward-thinking, engineering-led approach to system stability, shifting the paradigm from reactive firefighting to proactive safeguarding.

Capability	Team	Role Responsible
Pre-mortem and What-If Scenarios	Development	Tech Lead or Site Reliability Engineer (SRE)
Chaos Engineering & Testing	Development	Site Reliability Engineer (SRE) or Chaos Engineer
Synthetic Monitoring	Operations	Observability Engineer or SRE
Automated Risk Assessment in Change Management	Change & Release	Change Analyst or Release Manager
Release Strategy (Blue/Green, Feature Flags)	Development	Release Engineer or DevOps Engineer
Critical User Journey Identification	Development	Product Manager or UX Lead
SLIs & SLOs for Golden Signals with Error Budget	Development / Operations	SRE or Platform Engineer
Monitoring & Observability	Operations	Observability Engineer or Platform Engineer

Figure 2-6. *Capability and Role Responsible Mapping Example for proactive reliability measures (prevention)*

Reactive and Responsive Practices (Solving Incidents)

These capabilities are crucial for streamlining the process of managing and responding to critical system incidents effectively, ensuring rapid resolution and minimizing their negative impact on users and business operations. They are about organized, swift action when prevention fails.

Capability	Team	Role Responsible
Incident Management & Rotations	Development / Operations	On-Call Engineer (rotating SRE/Dev role)
On-Call Management	Operations	Incident Commander or SRE Manager
Patching, Endpoint Security	Operations	IT Security Engineer or Systems Administrator
Cloud Backup / Disaster Recovery	Operations	Cloud Infrastructure Engineer or DR Lead
Runbook Automation	Development / Operations	Platform Engineer or DevOps Engineer
Drift Detection & Reporting	Operations	Configuration Manager or SRE

Figure 2-7. *Capability and Role Responsible Mapping Example for reactive and responsive practices (solving incidents)*

Foundational Elements for Reliability (Improving Stability)

These capabilities represent the fundamental building blocks and enduring practices required for constructing, maintaining, and continuously enhancing reliable and scalable systems. They are the engineering bedrock upon which sustained stability is built.

Capability	Team	Role Responsible
Toil Management	Development	SRE or Platform Engineer
Resilience Design Patterns	Development	Software Architect or Tech Lead
Configuration Management	Operations	DevOps Engineer or Configuration Manager
Capacity Management	Operations	Capacity Planner or Infrastructure Engineer
Feature Flags	Development	Software Engineer or Feature Manager
Reliable Experiments	Development	QA Engineer or Experimentation Analyst
Automated Infrastructure Provisioning	Operations	Infrastructure as Code Engineer or Cloud Engineer
Infrastructure Utilization	Operations	FinOps Analyst or Cloud Operations Engineer
Continuous CI/CD	Development	CI/CD Engineer or DevOps Engineer
Cloud Cost Management	Operations	FinOps Lead or Cloud Cost Analyst

Figure 2-8. *Capability and Role Responsible Mapping Example for foundational elements for reliability (improving stability)*

Organizational and Cultural Aspects (Continuous Improvement)

These capabilities emphasize the indispensable importance of cultural shifts and a commitment to continuous learning within the organization. They are crucial for fostering a truly resilient, adaptive, and high-performing SRE environment where improvement is systemic and ingrained.

Capability	Team	Role Responsible
Blameless Post-Mortems	Development	Incident Reviewer (often SRE, Engineering Manager)
Psychological Safety	Organization-wide	Engineering Manager or Agile Coach
Build-Run Teams	Organizational Design	Head of Engineering or VP of Technology

Figure 2-9. *Capability and Role Responsible Mapping Example for organizational and cultural aspects (continuous improvement)*

To succinctly summarize the accountability and roles in the context of SRE: **Development teams** bear primary responsibility for proactive, design-focused reliability work. This critical function is typically spearheaded by tech leads, dedicated SREs, architects, and the software engineers themselves, who embed reliability into the code from inception. **Tech Operations teams** are meticulously responsible for the ongoing health, comprehensive observability, and swift recovery of systems during incidents. These teams are generally led by operations engineers, SREs, or system administrators, acting as the custodians of production stability.

Change & Release teams operate in a more flexible capacity; they can either be tightly embedded within agile squads or function as a crucial bridge connecting disparate teams. Their core mandate centers on governance, ensuring safe and controlled deployments, and maintaining meticulous compliance. Their leadership typically falls to release managers or specialized change coordinators. Finally, it's paramount to acknowledge that in highly regulated industries, the principle of **clear separation of duties** is absolutely critical. This mandates a rigorous separation of responsibilities between those who write code, those who possess access to production systems, and those authorized to activate new code in production environments, thereby mitigating risk and ensuring auditability.

Traditional Approach

In the **traditional approach** to IT operations, **Tech Operations** holds singular responsibility for managing the production environment. This traditional structure is typically layered into a hierarchy of **1st Level, 2nd Level, and 3rd Level support**.

The **1st Level** functions as the initial point of contact, often operating as a help desk, literally picking up the telephone when an end-user encounters an incident. Their primary task is to meticulously document the incident in a ticket and attempt to resolve it by following standardized operating procedures (SOPs), which are concise guidelines designed to address common issues. Most 1st Level teams are allotted no more than 20 minutes to solve an incident. If a resolution cannot be found within this timeframe, the ticket is escalated and transferred to **2nd Level support**.

The **2nd Level** possesses a broader array of capabilities, including more advanced technical knowledge, access to administrator user interfaces, and direct access to databases, along with a more extended timeframe for resolution. However, should they also be unable to resolve the issue, the 2nd Level further escalates the ticket to the **3rd Level**. Typically, only the 3rd Level team has access to the source code, enabling them to perform a deep root cause analysis of the problem. Once the system is restored to operation, but the underlying root cause remains unaddressed, a **problem ticket** is opened. This problem ticket allows for a more comprehensive investigation, providing ample time to delve into the issue and involve additional stakeholders, such as external vendors, other upstream or downstream application teams, or infrastructure teams. Ultimately, when a bug is identified and the source code is modified for deployment to production, a **change ticket** is created. This change ticket serves as critical documentation, meticulously recording precisely when, by whom, and why a specific modification was introduced into the production environment. This meticulous record-keeping is indispensable for audits, ensures transparency, and facilitates future bug-fixing efforts.

In this traditional organizational structure, the SRE team, or individuals acting in an SRE capacity, would most effectively connect with both the **3rd-level team and the Problem Management team**. This is due to their inherent overlapping functions; in many organizations, these two teams are often conceptually, if not formally, considered to be the same entity, as both are deeply involved in root cause analysis and systemic issue resolution. The SREs would also enhance the change and release management capabilities and support the developers to write robust code. The SRE influence here would be to infuse proactive reliability practices and blameless post-mortem learning into these critical reactive functions.

Build-Run: A Paradigm Shift in Technology Delivery

In stark contrast to the traditional IT model, the **build-run approach** represents a profound shift—a reimagining of how technology work gets done. Here, the very engineers who write the code also carry the torch of ensuring it runs flawlessly in production. It's like the architects of a bridge being the same people who monitor it daily for cracks and stress, racing to shore up weak spots before disaster strikes.

These are inherently cross-functional teams, typically about ten people strong, operating with remarkable autonomy. Within this model, ownership is holistic. They shepherd the entire lifecycle: writing the code, rigorously testing it, deploying it smoothly through a disciplined DevOps pipeline, and—critically—standing on the front lines when customer-impacting incidents occur. This model breeds a level of accountability and pride in operational excellence that's difficult to achieve in traditional setups.

While build-run teams are designed to be largely self-sufficient, they often receive lightweight support from a helpdesk that handles initial customer calls—frequently in multiple languages—and swiftly creates tickets routed to the right team. This keeps customer concerns from falling through the cracks while preserving the build-run team's focus.

A hallmark of high-functioning build-run organizations is how they **rotate production responsibilities**. Typically, two out of the ten engineers are on call each week. They own incident management, lead troubleshooting efforts, and vigilantly monitor the health of the systems. After a week, the baton passes seamlessly to two other engineers. This cadence ensures everyone maintains a deep connection to the operational realities of the product, prevents burnout, and democratizes critical knowledge.

Think of it as a rotating lighthouse crew—everyone takes turns keeping watch, maintaining the beacon, and steering ships clear of hidden shoals.

- **Diverse Skill Sets, Powerful Synergy:**

   Each team member brings a distinct strength to the table. One engineer might be an observability ace, deeply fluent in telemetry and tracing, while another could be a rock-solid systems thinker who excels at architecting fault-tolerant flows, and yet another a hawk-eyed tester. This tapestry of skills means the team doesn't just solve problems—they do so holistically, often uncovering root causes that would slip through in siloed models.

- **Slashed Communication Overhead:**

  Traditional structures fracture responsibilities into rigid silos: developers here, testers there, support somewhere else—like three separate bands trying to play a symphony without a conductor. This leads to handoffs that are slow, error-prone, and frustrating. Build-run teams obliterate these walls. Communication becomes immediate, handoffs are internal and almost invisible, and problem-solving accelerates dramatically.

- **Quality Baked In, Fewer Incidents:**

  Perhaps the most compelling advantage is that the same engineers who architect, build, and test the system are also on the hook to fix it when it breaks. They feel the operational pain firsthand. This naturally incentivizes them to build sturdier, more resilient systems from the start. It's the equivalent of having chefs wash the dishes—suddenly, they care a lot more about not burning sauce onto the pan.

A cornerstone of modern SRE practice is **enabling teams to evolve from traditional split models into full build-run structures**. This often involves coaching teams on how to integrate operational concerns into their design decisions, guiding them on observability best practices, and helping them build robust incident response playbooks.

However, it's important to recognize that for some teams—especially those constrained by regulatory or legal boundaries—this transition may not be feasible. In heavily regulated environments, there can be mandated separations between development and operations to satisfy compliance requirements.

Where build-run teams do take root, they are often clear indicators of a tech organization operating at a higher level of maturity. By internalizing both speed and stability goals within the same team, build-run structures dramatically reduce the classic tension between shipping fast and running reliably. They embody a balanced, disciplined approach that integrates change velocity with operational resilience—two sides of the same coin.

When you see these teams thriving, it's often a sign that the broader organization has cracked the code on marrying agility with durability. They've moved beyond the old game of lobbing issues over the wall, building instead a culture where everyone is invested in both innovation and steady hands on the wheel.

Returning to the lighthouse metaphor: in traditional setups, one crew might design the lighthouse, another might build it, and yet another entirely might be responsible for manning the light. Coordination gaps and finger-pointing abound when storms hit. In the build-run world, the same crew designs, constructs, maintains, and keeps watch from the tower. Their fingerprints are on every brick and every beam, so when ominous clouds gather, they don't scramble to figure out what's wrong—they already know where to look, and they're ready to act.

This is the essence of modern resilience: ownership that runs so deep, it naturally drives systems to be better from the start.

Processes and Workflows

The efficacy of an SRE-infused organization hinges on the seamless integration of its people, their roles, and their responsibilities into well-defined processes and workflows. Two critical areas demand meticulous implementation from the outset: the **software development process** and the **operating software process**.

Let's begin with the **software development process**. When planning a sprint or a development cycle, it is no longer sufficient to merely validate whether new features deliver business value. A fundamental shift is to explicitly validate whether new features also **measurably improve the stability and reliability of the software**. A best practice in this regard is to establish a clear, quantifiable target for reliability improvements within each sprint. For example, the sprint goal might not just be to ship new customer features but also to implement a specific number of resilience design patterns, ensuring that stability is a co-equal objective. Beyond feature development, the software development process must rigorously incorporate a diverse array of reliability tests, including sophisticated cybersecurity testing, comprehensive non-functional testing (performance, scalability, stress), and, critically, **chaos engineering experiments**, which proactively test system resilience under controlled failure conditions.

The second critical aspect is the process of **operating software**. This encompasses a suite of interconnected activities such as robust incident management, meticulous capacity planning, comprehensive service level management, and the pivotal postmortem process. For instance, during the crucial postmortem stage following an incident, the objective extends beyond mere incident resolution. We aim to precisely identify which resilience design patterns could have prevented or mitigated the incident and, more broadly, to maximize the return on investment (ROI) from every incident for the entire enterprise. By rigorously analyzing real-world incidents, we can significantly enhance the relevance and effectiveness of our proactive chaos experiments, directly informing future reliability investments. This cyclical feedback loop leads to a continuous improvement process, ensuring that the organization systematically learns and evolves with every operational challenge.

Within these established tech operations and software development teams, which are often already "well dialed in," an SRE assumes responsibility for a key element within each process step, directly impacting the overall reliability of the system. This integration ensures that SRE principles are woven into the daily fabric of engineering and operations.

Team	Process Step	Key SRE Responsibility
Tech Operations	**Incident Management**	• Triage and resolve high-severity incidents. • Participate in on-call rotations, embodying the shared responsibility for operational health. • Lead incident response coordination, including critical communications and war room facilitation. • Ensure the existence of comprehensive and actionable runbooks and playbooks for efficient resolution. • Automate detection and response mechanisms for recurring issues, reducing manual toil.
	Problem Management	• Conduct blameless post-incident reviews, fostering a culture of learning. • Analyze incident trends and recurring failure patterns to identify systemic weaknesses. • Propose and actively implement systemic fixes to address root causes. • Collaborate intensely with engineering teams on long-term reliability improvements.
	Change Management	• Rigorously evaluate the risk profile and blast radius of proposed changes. • Automate change risk assessment to enhance consistency and speed. • Monitor key signals meticulously during and after change rollouts to detect anomalies. • Define robust guardrails for safe deployments, preventing adverse impacts.
	Release & Deployment Management	• Enable the adoption of advanced release strategies like blue/green and canary deployments. • Ensure that rollbacks are not only possible but also safe and exceptionally fast. • Automate post-deployment verification, including smoke tests and health checks.
	Monitoring & Observability	• Define and meticulously implement service-level indicators (SLIs) that directly reflect user experience. • Maintain and optimize dashboards, alerts, and log aggregation systems. • Fine-tune alerting mechanisms to significantly reduce noise and mitigate alert fatigue. • Enable comprehensive tracing, logging, and metrics integration into all services. • Crucially, align error budget utilization with business stakeholders to balance features and stability.

Figure 2-10. *Process Step and Key SRE Responsibility Mapping Example for tech operations team*

Team	Process Step	Key SRE Responsibility
Software development	Requirements & Planning	• Advocate vociferously for reliability features to be prioritized in the backlog. • Define stringent reliability priorities and establish clear backlog thresholds. • Define comprehensive non-functional requirements, including availability, latency, and throughput. • Drive error budgets and complex risk tradeoff decisions, making reliability a business discussion.
	Design & Architecture	• Profoundly influence system design for optimal reliability, scalability, and operability. • Rigorously review architecture for inherent fault-tolerance and graceful degradation capabilities. • Provide expert advice on the judicious use of proven reliability patterns (e.g., circuit breakers, retries).
	Development & Implementation	• Support developers with essential tooling, such as test harnesses and chaos engineering tools. • Help enforce code standards that directly improve system reliability and maintainability. • Systematically inject comprehensive observability into the application code (metrics, traces, logs).
	Testing & Quality Assurance	• Develop sophisticated chaos tests and fault injection strategies. • Ensure that resilience and performance tests are an integral part of the continuous integration (CI) pipeline. • Validate the expected alerting and monitoring behaviors meticulously pre-release.
	Deployment	• Provide expert advice on establishing optimal thresholds for reliable deployment pipelines. • Advise on robust rollback strategies and automated testing within the deployment process. • Enforce critical deployment gates using service-level metrics, preventing unreliable code from reaching production. • Monitor the real-time impact of new releases on overall system health.

Figure 2-11. *Process Step and Key SRE Responsibility Mapping Example for software development team*

Technology and Tools

When embarking on the strategic imperative of improving technology within an organization, the immediate mental images that typically surface are often tangible assets: perhaps significant infrastructure upgrades, the adoption of shiny new tools, a radical redesign of data centers, or the implementation of cutting-edge software solutions. While these elements are undeniably important and form critical components of any modern tech strategy, it is crucial to recognize that they represent only a partial—and often superficial—component of the deeper, more profound narrative when it comes to true resilience and enduring stability. The tools are a means to an end, not the end itself.

Consider a practical, yet frustratingly common, scenario: a high-performing team is meticulously preparing for a looming peak season, perhaps a major holiday shopping surge or an annual financial reporting deadline. They have invested heavily in stress-testing their systems, diligently identifying areas where performance predictably degrades under anticipated load. In response, they smartly decide to enhance redundancy—adding robust failovers, optimizing complex routing algorithms, or introducing dynamic auto-scaling mechanisms to gracefully handle sudden, unpredictable traffic spikes. These are precisely the kind of intelligent, proactive moves that define engineering excellence.

However, the team then encounters an entirely unexpected, yet crippling, roadblock. The legacy monitoring tool they rely upon, once sufficient, simply cannot handle the sheer scale of the new environment, providing insufficient visibility or, worse, collapsing under the very load it's meant to monitor. And to compound the problem, a much-needed, modern replacement tool—one capable of providing the comprehensive observability required—is stuck in a bureaucratic quagmire, awaiting final approval. Suddenly, what appeared to be a straightforward, technical optimization spirals into a protracted, agonizing delay, necessitating a six-month procurement cycle, coupled with an arduous security review process. This isn't a failure of engineering; it's a failure of the overarching operating model and its ability to provision necessary technological enablers.

This kind of systemic delay is, regrettably, far from uncommon. The strategic imperative of getting the **right tooling in place—and doing so early and with deliberate intention**—is absolutely crucial for the success of any reliability initiative. When an application fundamentally lacks essential observability features or robust resilience testing capabilities from its inception, it becomes almost impossibly difficult—and prohibitively expensive—to retrospectively fix these critical gaps when systems are already under immense stress, performing under peak load, or, worse, experiencing a live incident. In many enterprises, the process of onboarding a seemingly innocuous new tool involves navigating a Byzantine, protracted process encompassing extensive vendor screening, multiple layers of internal approvals, and a painstaking production readiness review. This labyrinthine journey consumes precious time—time that no one can afford to lose, especially during a crisis where minutes translate directly into millions in lost revenue or irreversible reputational damage.

For specialized practices like site reliability engineering, having a dedicated suite of specialized tools—such as those custom-designed for precisely defining Service Level Objectives (SLOs) or for systematically conducting complex Chaos Engineering experiments—isn't merely a "nice to have" luxury. It is, in fact, an **absolute essential requirement**, forming the very backbone of their operational effectiveness. These indispensable tools empower teams to meticulously simulate various failure scenarios in highly controlled environments, gain profound insights into a system's breaking points, and systematically improve its resilience *without* inadvertently causing real-world outages or customer impact. Crucially, robust safety mechanisms are inherently built into these tools, rigorously ensuring that these proactive tests never trigger unintended cascading failures in production. This transforms the act of breaking things into a structured, safe, and profoundly educational exercise.

Beyond these specialized instruments, there is the foundational importance of **monitoring and observability**—the equivalent of a sophisticated radar system for every component running in production. With the right instrumentation consistently applied and a mature, deeply ingrained understanding of how to interpret the complex tapestry of signals emanating from the system, teams are empowered to detect subtle abnormal behaviors long before any impact is felt or even noticed by end-users. In this context, a meticulously tuned and intelligently routed alert can often represent the decisive difference between a minor, easily addressable hiccup and a full-blown, catastrophic incident that brings operations to a standstill. It's the early warning system that allows for proactive intervention.

The profound truth is that resilience does not magically begin when something inevitably goes wrong. It originates far earlier in the lifecycle—it is meticulously designed into the architecture, meticulously planned at the strategic level, and built into the very tools and platforms we choose to trust. And, more often than not, the most formidable challenges encountered in achieving this resilience are not purely technical; they are, at their core, organizational and cultural. This underscores the holistic nature of SRE transformation.

SRE Capabilities and Culture

We have already established "Reliability" as our top-level strategic domain—our foundational Level 1 capability. Now, it is imperative to delve deeper into the intricate fabric of the Level 2 and, more specifically, the granular Level 3 capabilities that are indispensable for truly strengthening and sustaining reliability across the enterprise. When we speak of "people" within the context of SRE, it encompasses far more than merely skills and training, although these aspects are undeniably critical. The intrinsic technical strength and the collective acumen of your teams are paramount.

If you are embarking on your SRE journey from a nascent stage, the logical starting point is a meticulous assessment of your existing teams' current capabilities and readiness. It is highly probable that you will discover some individuals or teams already organically moving in the right direction, possessing an intuitive blend of the requisite skills and a foundational understanding. Perhaps they have imported this invaluable expertise from prior professional experiences at other companies, or they may have cultivated it independently through an innate personal passion for operational excellence and systems thinking. These pioneering teams are not merely early adopters; they become the vital "nucleus" of your burgeoning SRE practice. They inherently grasp both the technical intricacies and the nuanced business hurdles unique to your specific company, and they already possess a nascent understanding of what SRE truly entails. These individuals and teams are your internal champions, capable of seeding the transformation.

Working collaboratively with this nascent SRE nucleus, the next strategic step is to precisely define the necessary training needs. This involves meticulously identifying specific knowledge gaps and subsequently providing targeted, robust training programs for your new capabilities, such as the critical discipline of chaos engineering. Furthermore, depending on your chosen tooling ecosystem, many vendors often provide specialized certifications and comprehensive training pathways, which can serve as valuable external validation and skill-building resources.

Beyond individual skills, the subsequent, equally crucial step is to strategically bring your teams together and diligently foster pervasive, high-quality communication. **Inter-team communication** specifically refers to ensuring consistently effective and transparent dialogue among all your relevant engineering teams—development, operations, security, and product. The objective is to cultivate a vibrant **community of practice**, a collaborative ecosystem designed for the fluid sharing of information, the dissemination of best practices, and the nurturing of a profound sense of belonging among practitioners. This community should be underpinned by ready access to shared reporting, comprehensive documentation, and practical code snippets, facilitating learning and acceleration. To achieve this, it is imperative to establish clear, accessible channels for both **reporting** and **documenting**, for instance, the precise results and critical insights gleaned from chaos experiments, ensuring that every learning opportunity is captured and shared systematically.

One of the most profoundly important, yet often overlooked, aspects of constructing a robust and enduring SRE practice is not solely contingent upon the sophisticated tools we deploy. More fundamentally, it resides in **how we collectively think about reliability, stability, and risk** across the entire organizational landscape. This represents a profound cultural shift that underpins all technical implementations.

It commences with a fundamental change in mindset. The ambition is not merely for individual teams to learn in isolation from incidents or outages; rather, the overarching goal is for the **entire organization to learn, adapt, and continually evolve** from these experiences. This requires a deep, shared understanding that every enterprise possesses its own unique and inherent appetite for risk, and consequently, that reliability is not a monolithic, one-size-fits-all concept. The level of acceptable risk varies dramatically depending on context, industry, and strategic objectives as described above in this chapter.

But understanding risk extends beyond just abstract systems; it is also profoundly about **how people behave when things inevitably go wrong**.

The second critical facet of fostering true resilience is inherently tied to **how we collectively respond to failure**. When an incident strikes, do teams naturally collaborate seamlessly to dissect and fix the root cause, driven by a shared mission to improve the system? Or do they instinctively scramble to protect individual interests, to assign blame, or to deflect responsibility? Does the organization, at its core, treat failures as invaluable learning opportunities, catalysts for systemic improvement? Or does it relentlessly pursue someone to blame, initiating punitive measures? These subtle yet powerful cultural signals are the definitive indicators of whether a company truly embodies and embraces a genuine mindset of resilience.

Organizations that actively encourage learning over punishment—where the practice of **blameless postmortems** is not just an ideal but an ingrained norm, and where significant system improvements systematically follow every major incident—tend, over time, to construct stronger, more inherently adaptive operations. In vivid contrast, a pervasive culture characterized by finger-pointing, fear of retribution, or a reluctance to admit mistakes can tragically make teams hesitant to surface issues early, even when they are minor. This suppression of transparency leads to deeper, more insidious risks and, inevitably, significantly longer downtimes when problems finally do erupt.

This is precisely why risk tolerance is not confined to being merely a technical concept. It is, in fact, **deeply intertwined with the very fabric of organizational culture**. To advance strategically, we must embark on a dual assessment: evaluating not only the robustness of our systems but, equally critically, the adaptability and integrity of our collective behaviors. Because, ultimately, the precise manner in which we handle failure today profoundly shapes our intrinsic ability to not only survive but truly thrive tomorrow.

Keeping systems reliably operational at scale transcends mere reactive fixes; it fundamentally demands a proactive commitment to establishing the right habits, deploying appropriate tools, and fostering optimal ways of working. Over years of observation across diverse teams and industries, distinct patterns of success have consistently emerged. Certain key capabilities stand out as the absolutely essential building blocks for implementing site reliability engineering (SRE) effectively within real-world enterprises. These capabilities encompass a broad spectrum of expertise— engineering, architectural design, and operational acumen—all converging on core principles such as pervasive automation, cultivating a healthy team culture, and architecting systems to gracefully handle and recover from failures. Together, these building blocks empower teams to accelerate innovation, consistently maintain resilience, and, critically, to extract maximum learning from every single incident. For clarity, these building blocks can be grouped thematically to better understand their specific purpose and strategic contribution:

Proactive Reliability Measures (Prevention)

These capabilities are the essence of a forward-thinking SRE approach, meticulously designed to anticipate and prevent issues long before they have the chance to negatively impact users. They embody a strategic shift from reactive firefighting to engineered foresight, ensuring system stability is built in, not bolted on.

- **Pre-mortem and What-If Scenarios:** This critical practice involves engaging in proactive, structured scenario planning to systematically identify potential failure modes and devise effective mitigation strategies *before* incidents ever occur. This foresight significantly enhances overall system resilience and facilitates the early identification of latent vulnerabilities, transforming potential weaknesses into recognized, managed risks.

- **Chaos Engineering & Testing:** This advanced discipline involves the deliberate, controlled introduction of failures into a system to rigorously test its integrity, observe its response mechanisms, and validate its resilience. By intentionally creating disruptions, teams can empirically validate system robustness, pinpoint unforeseen weaknesses, and prepare comprehensively for real-world failure scenarios, building confidence in system behavior under duress.

- **Synthetic Monitoring:** This capability focuses on proactively testing and continuously monitoring application performance by simulating realistic user interactions through automated scripts. It ensures continuous checks on system responsiveness, availability, and functionality, enabling the early detection of issues, often before actual users are impacted.

- **Automated Risk Assessment in Change Management:** This leverages the power of AI and advanced analytics to provide data-driven insights into the potential impacts of proposed system changes. It objectively evaluates modifications, predicts potential risks, and offers actionable recommendations to minimize unintended consequences, embedding intelligence into change processes.

- **Release Strategy:** This focuses on fundamental techniques designed to significantly reduce downtime and mitigate deployment risks. Through advanced strategies like blue/green deployments and feature flags, the aim is to minimize disruptive weekend deployments and alleviate team workloads, promoting safer, more controlled releases.

- **Critical User Journey Identification:** This involves the meticulous process of identifying and prioritizing the most essential user workflows that directly underpin business value and crucial user experience. By focusing engineering efforts on these high-impact areas, resources are directed where they matter most, maximizing the return on reliability investments.

- **SLIs & SLOs for Golden Signals with Error Budget:** Customer experience is precisely measured through Service Level Indicators (SLIs), which capture granular data on key metrics such as latency, error rates, and data quality. Based on these SLIs, Service Level Objectives (SLOs) are established to define desired reliability targets. Teams then manage Error Budgets, providing a powerful, quantitative approach to balancing the continuous drive for new features with the non-negotiable demand for reliability by defining a permissible amount of unreliability within a given period.

- **Monitoring & Observability:** This encompasses the comprehensive monitoring of all business-critical functions through systematic logging, distributed tracing, and meticulous metrics collection. Observability, a deeper concept than mere monitoring, enables a profound understanding of a system's internal state by rigorously examining its external outputs, facilitating faster troubleshooting and data-driven decision-making, and transforming raw data into actionable insights.

Reactive and Responsive Practices (Solving Incidents)

These capabilities are designed to streamline the process of managing and responding to critical system incidents with maximum efficiency. Their focus is on ensuring rapid resolution, minimizing impact, and restoring service, acting as the tactical arm of SRE when prevention has been breached.

- **Incident Management & Rotations:** Incident resolution is systematically addressed through a **build-run rotation model**, where development engineers rotate into on-call roles, balancing their primary implementation work with a critical sense of operational empathy. Teams assign members weekly for production support, dedicating their focus exclusively to incident resolution during their rotation, fostering shared ownership of operational health.

- **On-Call Management:** Structured and effective on-call management is absolutely essential for proficient incident handling. This often utilizes established frameworks like the Incident Command System to systematically manage incidents, effectively communicate with diverse stakeholders, and resolve issues within a clear organizational structure. Automated on-call support tools further streamline the process, intelligently routing incidents to the appropriate teams and escalating based on severity thresholds, thereby minimizing response time and mitigating alert fatigue.

- **Patching, Endpoint Security:** This refers to the continuous, diligent practice of applying security updates and maintaining robust security postures at all system endpoints. Its purpose is to proactively prevent vulnerabilities from being exploited and to ensure the unwavering integrity of all systems, acting as a crucial line of defense.

- **Cloud Backup/Disaster Recovery:** This involves the meticulous establishment of comprehensive plans and resilient mechanisms for reliably backing up critical data and enabling swift system recovery in the unfortunate event of a catastrophic disruption or failure. This ensures business continuity even in the face of major unforeseen events.

- **Runbook Automation:** This capability focuses on automating scripts that provide predefined, systematic steps to help solve common incidents. These automated actions can be initiated from various devices, significantly reducing manual intervention, accelerating resolution times, and ensuring consistent incident handling.

- **Drift Detection & Reporting:** This capability is centered on systematically identifying and reporting any deviations from desired or standardized system configurations. It plays a crucial role in maintaining consistency across environments, preventing unexpected behaviors, and ensuring that systems adhere to their intended design, thereby reducing the likelihood of configuration-induced incidents.

Foundational Elements for Reliability (Improving Stability)

These capabilities constitute the essential building blocks and core practices for meticulously constructing and maintaining highly reliable, scalable, and performant systems. They are the engineering pillars that ensure long-term stability and robust operation.

- **Toil Management:** This involves the strategic identification and proactive reduction of repetitive, manual, and low-value operational tasks. By leveraging automation and self-service tools, organizations can significantly increase team productivity, liberating engineers to focus their valuable time and expertise on more strategic, high-impact work, driving innovation rather than drudgery.

- **Resilience Design Patterns:** These are battle-tested architectural approaches that fundamentally enhance system reliability and fault tolerance. By implementing strategies such as circuit breakers (to prevent cascading failures), retry mechanisms (to handle transient errors gracefully), and timeouts (to prevent indefinite waits), these patterns improve overall system stability and proactively prevent small failures from escalating into widespread issues.

- **Configuration Management:** This capability ensures consistent, version-controlled, and fully reproducible system configurations. By managing infrastructure and application configurations as code, it significantly reduces configuration "drift" across environments and enables easier, more reliable system reproducibility, leading to predictable deployments.

- **Capacity Management:** This involves predicting, continuously monitoring, and strategically optimizing system resources to consistently meet performance and scalability requirements. It entails analyzing usage patterns and dynamically adjusting infrastructure to maintain peak performance while simultaneously controlling costs, ensuring resources are optimally utilized.

- **Feature Flags:** These are powerful software development techniques that enable controlled rollout and testing of new features. By allowing developers to dynamically toggle functionality on or off in production, feature flags significantly reduce deployment risks and support highly flexible, iterative software delivery, enabling A/B testing and dark launches.

- **Reliable Experiments:** This refers to the systematic and rigorous testing of system changes with tightly controlled variables and precisely measurable outcomes. This disciplined approach allows for the validation of hypotheses and a significant reduction of uncertainty in system modifications, ensuring that changes lead to desired improvements.

- **Automated Infrastructure Provisioning:** This capability ensures the automatic setup and configuration of all necessary infrastructure components. It dramatically reduces manual effort, accelerates environment creation, and significantly increases consistency in deployments, paving the way for true infrastructure as code.

- **Infrastructure Utilization:** This focuses on actively optimizing the efficient use of existing computing resources. By maximizing the utility of deployed infrastructure, organizations can improve cost-effectiveness, enhance performance, and defer unnecessary hardware or cloud spending.

- **Continuous CI/CD:** Continuous Integration and Continuous Delivery (CI/CD) frameworks are paramount for ensuring that code changes are delivered rapidly, yet with unwavering safety. They incorporate built-in validation gates that meticulously catch regressions and other issues before they ever reach production, acting as an automated quality control system.

- **Cloud Cost Management:** This capability focuses on actively managing and rigorously optimizing expenditures related to cloud services. It ensures efficient resource allocation, prevents runaway costs, and aligns cloud spending with business value, treating cloud resources as a financial asset to be optimized.

Organizational and Cultural Aspects (Continuous Improvement)

These capabilities fundamentally underscore the critical importance of cultural shifts and a deeply embedded commitment to continuous learning. They are essential for fostering a resilient, adaptive, and truly innovative SRE environment where mistakes become lessons and teams feel empowered to continuously evolve.

- **Blameless Post-Mortems:** These are non-punitive, forward-looking reviews of system incidents, with the singular focus on learning and systemic improvement. Teams conduct thorough, objective analyses of incidents, rigorously emphasizing systemic issues, process flaws, and environmental factors rather than assigning individual blame. This practice profoundly promotes a culture of continuous learning and directly drives meaningful, long-term system improvements.

- **Psychological Safety:** This involves the deliberate creation of an environment where team members feel intrinsically safe to take calculated risks, openly share their ideas, and, crucially, admit mistakes without fear of retribution or negative consequences. This fosters a powerful culture of trust, open communication, continuous learning, and robust innovation, where vulnerabilities are opportunities for growth.

- **Build-Run Teams:** This transformative organizational structure integrates development and operations functions, systematically promoting shared responsibility and holistic end-to-end ownership for software products. Cross-functional teams are jointly accountable for both building and maintaining software systems, leading to dramatically improved collaboration, faster feedback loops, and a profound alignment of development priorities with operational realities and demands.

Performance Metrics: Measuring What Matters

In this pivotal section, we will precisely define how our new Reliability domain strategically connects with the Chief Information Officer (CIO) and other vital key stakeholders across the enterprise. When initiating any new, transformative capability, obtaining unwavering support and explicit commitment from leadership is not merely beneficial; it is absolutely crucial for long-term success. They must possess a profound understanding of the intrinsic value that SRE brings to the organization and clearly perceive how it seamlessly integrates into the existing enterprise structure, reinforcing its strategic alignment. This executive buy-in directly correlates with the effectiveness of the processes you establish. By meticulously assigning people who possess the right blend of skills, a growth-oriented mindset, and unwavering dedication, you will cultivate enduring capabilities that are built to last. We typically advocate for initiating these efforts quickly, often with a small, highly focused pilot team, to demonstrate early value and build momentum.

However, to sustain leadership buy-in well beyond the initial enthusiasm and the first budget allocation, it is imperative to consistently track your success with clear, measurable performance indicators. This allows you to compellingly showcase the tangible impact of your SRE investments. Beginning with the second budget round, the ability to robustly justify the sustained effort and resources invested becomes a critical requirement. This leads directly to the necessity of a well-defined reporting structure. Your SRE team must meticulously share reports frequently across the organization, ensuring that all efforts are transparently aligned with overarching business goals and the broader organizational strategic initiatives. This can be effectively achieved by focusing your reporting on applications or products that are demonstrably critical to your customers, directly linking reliability to customer satisfaction and business outcomes.

For site reliability engineering to truly thrive and deliver its full value, teams require the right metrics—metrics that extend far beyond simplistic uptime percentages. They need signals that accurately reflect the nuanced user experience and provide clear guidance for prioritizing work. Our experience consistently reveals that most companies, when first approaching reliability, tend to focus exclusively on singular reliability measurements. However, in a complex, interconnected environment, reliability cannot exist in isolation; it does not stand on its own.

Consider a thought experiment: I could theoretically cease all changes from being deployed into production. This action would, without doubt, dramatically improve my reliability metrics, as the system would be static and potentially highly stable. I could then dedicate all resources to meticulously improving and fine-tuning my existing systems in production, ensuring maximal stability and robustness for my customers, but at the cost of zero new features. This extreme scenario vividly illustrates that we must consider multiple, interdependent dimensions when measuring true reliability. We need to balance these dimensions holistically to remain relevant and to optimize our entire approach for the ultimate benefit of our customers.

A truly resilient organization must strive for a multi-dimensional equilibrium. We need to be **robust** (reliable and stable), **fast** (able to deliver new features rapidly), foster **happy** customers and engaged employees, and achieve all of this in a **cost-efficient** manner. To achieve this delicate balance, our resiliency organization must rigorously set all measurements in a broader perspective. Even if we are not directly responsible for a particular measurement, such as deployment frequency, it serves as a crucial anchor point to make our core reliability measurements truly meaningful. Therefore, we must strategically include such complementary metrics in our overall reporting. Our key metrics, encompassing this holistic view, should include

Resiliency and Stability:

- **Mean Time to Restore (MTTR):** This crucial metric quantifies how quickly we can effectively recover from an outage, reflecting the efficiency of our incident response.

- **Change Failure Rate:** This metric tracks how often new deployments inadvertently introduce problems or cause failures, indicating the safety and quality of our release processes.

- **Error Budget Burn Rate:** This metric tracks how quickly we are consuming our agreed-upon "budget" for unreliability, providing a real-time indicator of whether we are staying within our predefined reliability targets.

- **Number of Critical Incidents:** This metric provides a high-level overview of the frequency of major disruptions, highlighting what is impacting customers most significantly.

- **System Uptime:** While not the sole metric, this still remains a fundamental measure of whether we are consistently meeting our Service Level Agreement (SLA) and Service Level Indicator (SLI) targets for availability.

Speed for New Features:

- **Deployment Frequency:** This metric measures how often we are successfully releasing new code or features to production, reflecting our agility and continuous delivery capabilities.

- **Lead Time for Changes:** This critical metric quantifies the duration from the initial code commit to its successful deployment in a production environment, indicating the overall efficiency and speed of our development pipeline.

Happy Customers and Employees:

- **Net Promoter Score (NPS):** This widely recognized metric assesses whether users would recommend our product or service to others, providing a powerful indicator of overall customer loyalty and satisfaction.

- **Customer Satisfaction (CSAT):** This directly measures user satisfaction with various aspects, including support interactions, overall experience, and product functionality, providing granular feedback on the customer journey.

- **Customer Retention & Churn:** These metrics track our ability to grow user engagement and retain our existing customer base, reflecting the long-term health of our user relationships.

- **Employee Satisfaction:** This internal metric gauges the happiness and engagement levels of our team members, often collected through methods like Employee Net Promoter Score (eNPS) or internal surveys, recognizing that internal well-being impacts external delivery.

Cost of Reliability:

- **Within Budget:** This fundamental financial metric assesses whether our infrastructure and operations costs are staying aligned with predefined forecasts and budget allocations.

- **Cost per Incident:** This metric quantifies the financial impact of each instance of downtime, including direct costs, lost revenue, and recovery expenses, providing a clear financial consequence of unreliability.

- **Engineering Time Spent on Reliability Work:** This metric tracks the proportion of engineering resources invested directly into resilience-related development and operational improvements, providing insight into whether we are wisely allocating effort towards long-term stability.

Organizations must internalize that the relentless pursuit of resiliency necessitates a delicate, continuous balancing act across multiple dimensions that extend far beyond the narrow confines of SRE and core technology. If one attempts to move too fast, the inherent risk of breaking things increases significantly. Conversely, an excessive focus solely on stability, without concurrent attention to speed, will inevitably cause progress to slow to a crawl. If cost-cutting measures are pursued too aggressively or too deeply, the organization sacrifices crucial flexibility and future investment capacity. Similarly, a singular chase for customer happiness without adequately supporting and nurturing your internal teams will almost certainly lead to pervasive burnout and diminished long-term effectiveness. The true, complex challenge lies in finding and maintaining a dynamic equilibrium among all these critical factors: speed of delivery, unwavering stability, optimized cost structures, genuinely happy users, and, fundamentally, healthy, engaged teams. To ensure this intricate balance is maintained, we must continuously report and manage CIO-level metrics that transcend pure resiliency, keeping all dimensions in check. This holistic approach begins and is reinforced at the individual team level.

These comprehensive metrics provide leaders with the crucial insights necessary to focus not merely on delivering *more* software but, critically, on delivering it *safely*.

From Firefighting to Prevention

In a traditional operational model, problems are almost invariably discovered only after they have already inflicted an impact on users. Teams then reactively scramble to respond, often operating without the benefit of sufficient, real-time data or the necessary tools. In this scenario, resilience devolves into a reactive effort—a perpetual state of solving yesterday's problem rather than proactively preventing tomorrow's. This is the classic "hero culture" of operations, where heroics are necessary because foundational issues persist.

The SRE model represents a profound and fundamental inversion of this reactive approach. Instead of waiting for disaster, **observability tools proactively surface anomalies and potential issues long before they escalate into full-blown outages**. **Chaos experiments deliberately expose weaknesses in a controlled environment** long before those vulnerabilities can be exploited by real users or unexpected events. **SLOs ensure that teams are precisely aiming for the right, optimal level of reliability**—neither over-engineering with unnecessary expense nor under-engineering with undue risk. And, critically, **blameless postmortems transform every system failure into an invaluable, collective learning opportunity**, fostering continuous improvement rather than a cycle of fear and blame.

By making a deliberate and strategic investment in proactive reliability, organizations can achieve a multitude of cascading benefits. Teams significantly reduce the financial and reputational cost of downtime, cultivate and deepen customer trust through consistent service, and fundamentally liberate themselves from reactive firefighting to confidently innovate and push the boundaries of what is possible. It is a strategic shift that moves technology from being a reactive cost center to a proactive competitive advantage.

The SRE Tech Operating Model is far more than a mere collection of processes or a framework of tools. It is, at its very core, a **fundamental mindset transformation**. It profoundly challenges teams to internalize and take absolute ownership of reliability from the very inception of a project—to treat it not as a regulatory checkbox or an operational burden but as an **intrinsic, core product feature**. When meticulously implemented and embraced, SRE transcends its technical definition to become the veritable "operating system" for modern, high-performing technology organizations. It empowers teams to move with unprecedented speed, yet with inherent safety. It enables businesses to confidently take bold, strategic bets, secure in the knowledge that their foundational systems are robust and resilient. Ultimately, SRE transforms resilience from a burdensome necessity into an undeniable, powerful competitive advantage, defining the enterprises that will thrive in the digital age.

Summary

SRE is not merely beneficial; it is an **essential strategic imperative** for modern organizations grappling with increasingly complex technology landscapes, rapidly escalating customer expectations, and a tightening web of regulatory demands. The traditional, bifurcated "build" and "run" models, characterized by fragmented responsibilities, have demonstrably led to disjointed efforts and persistent reliability issues. SRE offers a powerful, unified approach, meticulously bringing together engineering, operations, and business teams under a shared, singular focus: the unwavering delivery of reliable, scalable, and inherently trustworthy systems.

This chapter has systematically outlined five key, interconnected dimensions that collectively form the bedrock of a reliability-focused Tech Operating Model:

- **Organizational Structure and SRE Roles:** This dimension meticulously defines SRE's strategic placement and its inherent responsibilities within the enterprise, emphasizing its indispensable integration throughout the entire software lifecycle. A modern, resilient organizational structure is characterized by centralized accountability, ideally vested in a single individual or dedicated function reporting directly to the CIO. The precise implementation of SRE capabilities can manifest in various forms: as a dedicated, embedded role, as a pervasive skill set cultivated across all engineers, or consolidated within a central "SRE Hub."

- **Reliability-Driven Processes and Workflows:** This dimension encompasses the standardization of systematic methods for embedding and ensuring reliability across every phase of the software lifecycle, from initial planning and meticulous design through to seamless deployment and ongoing production operations.

SRE assumes a crucial, active role in incident management, proactive capacity planning, and the critical postmortem process, all designed to foster continuous learning and systemic improvement.

- **Reliability Technologies and Tooling:** This dimension underscores the critical importance of selecting and implementing the right foundational technology stack and specialized tools for achieving and sustaining reliability. This includes, but is not limited to, robust observability platforms, powerful automation frameworks, and advanced chaos engineering tools, all of which are essential enablers of SRE practices.

- **SRE Capabilities and Culture:** This dimension moves beyond mere technical skills to focus on cultivating the necessary aptitudes and, crucially, fostering a "blameless learning culture" within the organization. In such a culture, the entire enterprise learns and evolves systematically from incidents, transforming failures into valuable insights. A profound understanding of the organization's intrinsic risk tolerance is absolutely key to strategically tailoring and ensuring the successful adoption of SRE strategies.

- **Reliability Performance Metrics:** This dimension precisely defines how the success of SRE initiatives and the overall reliability of systems are quantitatively measured, tracked, and reported. The core of this measurement framework relies on Service Level Indicators (SLIs), Service Level Objectives (SLOs), and the strategic utilization of Error Budgets. However, a truly balanced approach to metrics transcends singular reliability measures, encompassing speed of delivery, holistic customer and employee satisfaction, and the optimized cost of reliability, ensuring a comprehensive view of operational health.

The chapter strongly reiterates that the existing Tech Operating Model should be thoughtfully adapted and augmented, rather than entirely rewritten from scratch. This adaptation must be meticulously aligned with the overarching business strategy, with a clear focus on optimizing the utilization of all technology resources to achieve strategic objectives.

Ultimately, the SRE Tech Operating Model initiates a fundamental transformation in how organizations fundamentally approach system reliability. By embedding reliability as a core, non-negotiable product feature and strategically shifting the organizational focus from reactive problem-solving to proactive prevention, SRE empowers businesses to innovate with unprecedented confidence. This transformative approach not only leads to a significant reduction in costly downtime and a substantial enhancement of customer trust, but it also elevates resilience from being a burdensome necessity to a crucial, strategic competitive advantage in the modern digital economy.

## CHAPTER 3

SRE in a Legacy Enterprise: Navigating Deep-Seated Challenges and Driving Transformative Change

This pivotal chapter delves into the very heart of site reliability engineering's (SRE) transformative power within the complex, often labyrinthine, world of the legacy enterprise. As established in Chapter 1, the digital age demands unprecedented speed, but without the inherent "guardrails" and "seatbelts" of reliability, this velocity becomes a liability, jeopardizing customer trust and organizational reputation. Chapter 2 further illuminated how traditional operating models, fractured by fragmented ownership and reactive approaches, crumble under the weight of escalating customer expectations, stringent regulatory demands, and relentless cybersecurity threats. This chapter builds directly on those foundational insights, offering a granular examination of the deeply entrenched challenges that legacy IT organizations face and, more importantly, providing a comprehensive blueprint for how SRE principles can not only reshape traditional IT operating models but also redefine the very essence of technological excellence and business value. It aims to provide a practical guide, balancing the compelling storytelling of a foundational text like Gene Kim's *The Phoenix Project* with the analytical precision and structural clarity found in works on enterprise architecture. This chapter delivers a deep, expert-level analysis woven with relatable examples and

© Florian Hoeppner, Francesco Sbaraglia 2025
F. Hoeppner and F. Sbaraglia, *Mastering Site Reliability Engineering in Enterprise*,
https://doi.org/10.1007/979-8-8688-1448-8_3

compelling insights, demonstrating how SRE moves beyond mere technical practice
to become a cornerstone of strategic business advantage. The narrative navigates the
enduring divide between traditional IT structures and modern demands, exposing
why legacy models inherently struggle to deliver resilience at speed. It delves into
the paradigm shift SRE instigates, particularly through the "You Build It, You Run It"
philosophy, which fundamentally re-architects team dynamics and accountability. A
significant portion is dedicated to modernizing operational processes—from incident
and problem management to the intricate dance of segregation of duties and IT
change management—showing how SRE's data-driven, proactive methodologies
supplant reactive firefighting. Finally, it explores how to quantify SRE's often-unseen
value, demonstrating its tangible benefits beyond mere uptime, encompassing speed,
customer satisfaction, and the vital well-being of engineering talent. This is a chapter
about transformation, about breaking free from the constraints of the past to build a
future where resilience is not an aspiration but an ingrained reality, driving competitive
advantage and ensuring enterprise longevity.

The Enduring Divide: Why Legacy IT Structures Struggle with Modern Demands

The operating model serves as the **centerpiece of any IT organization**, fundamentally
determining its success in developing and maintaining software. This model is not a
static blueprint but a dynamic reflection of how strategic business and IT objectives
are translated into tangible capabilities, team structures, and the intricate interactions
between these entities. As SRE transformation journeys across diverse enterprises
demonstrate, while the imperative to optimize for resilience is undeniable, the
starting point for most legacy organizations is rarely a blank slate. Instead, it's a
landscape shaped by decades of accumulated practices, deeply entrenched mindsets,
and fragmented responsibilities—a reality that, if not understood and addressed
with nuance, can significantly impede or even derail the most well-intentioned SRE
initiatives.

Consider, for instance, a large financial institution that has been operating for over
a century. Its IT strategy, articulated by the CIO, might proudly declare a mission to
"drive a market-leading IT service platform for global clients, reduce time-to-market
for new features, and deliver best-in-class, stable, and reliable customer service." While
this is a commendable aspiration, perfectly aligned with modern demands, beneath

this aspirational veneer often lies a deeply complex, often paradoxical, reality. The very structures designed for stability in a bygone era—hierarchical reporting lines, stringent separation of duties, and a project-centric funding model—now inadvertently stifle the very agility and resilience they were originally intended to protect. This chapter systematically unpacks the prevailing conditions within large enterprises, revealing the inherent contradictions and the systemic vulnerabilities that SRE is uniquely positioned to address.

The Chasm of Disconnectedness: How Decades of Dev Versus Ops Separation Hinders Speed and Reliability

A superficial glance at the job descriptions of CIOs or CTOs in large enterprises would suggest that resilience, stability, and reliability are core tenets of their roles. Indeed, these concepts are often formally enshrined within business and IT continuity plans, security frameworks, and operational structures designed to respond to outages, often leveraging multi-cloud strategies. Yet, if we probe deeper, asking for quantifiable indicators of a company's actual stability or resilience, the answers often become remarkably opaque and unsatisfactory.

The seemingly simple question, "How stable is our IT?", often receives a complicated response that only comes after the fact. Leaders might cite the **number of incident tickets or outages** as proxies for stability. However, an incident ticket is, by its very definition, a post-mortem indicator—a record of a problem that **has already occurred**. Its value lies in reaction, not prevention. Relying solely on such metrics is akin to a pilot assessing aircraft stability by reviewing crash reports *after* an accident; it tells you where you went wrong but offers little foresight.

Furthermore, the metrics often reflect human, rather than systemic, issues. The "fluctuation of IT people per location" might hint at underlying stress, but it doesn't directly measure system robustness. The truth is, many IT leaders "guess" at their organization's stability, relying on the frequency of escalations from business counterparts—escalations that, again, only occur **after the fact**, when it's already too late to prevent the disruption. They may recognize instability after it happens, but they lack the real-time visibility needed to anticipate and prevent the next disruption.

This pervasive lack of proactive, measurable insights is a critical vulnerability. The modern enterprise, as Chapter 1 underscored, is a "fractal and constantly under pressure" system. External forces, such as the COVID pandemic and its dramatic shift in work behavior, geopolitical conflicts like the war in Ukraine and its associated sanctions, or the sudden, resource-intensive demands of a new acquisition or a rapidly scaling product, all exert immense pressure on IT systems. These unforeseen "black swan" events, as Chapter 1 introduced, are increasingly frequent and closer than we expect. Without real-time, technical measurements—data-driven insights transparently displayed on dashboards—decisions about whether to push more change or delay deployments remain largely uninformed, reactive, and prone to error. **This is precisely the void that SRE is designed to fill. It fundamentally transforms how enterprises approach stability and resilience, shifting the focus from post-incident analysis to proactive prevention**. It demands a holistic adjustment of the IT operating model, encompassing new capabilities, refined processes, upskilled talent, and an optimized organizational structure.

At the genesis of any enterprise transformation, a deep, empathetic understanding of the existing "as-is" state is paramount. From extensive experience across numerous companies, a starkly consistent pattern emerges: the vast majority of legacy organizations are, to varying degrees, structurally cleaved into distinct "build" (development) and "run" (operations) organizations. This fundamental schism permeates every level, from individual teams all the way up to the C-suite, often manifesting as separate "Head of Development" and "Head of Operations" leadership teams.

This deeply ingrained separation isn't merely an organizational chart anomaly; it manifests in profound differences in daily work, influencing everything from the **understanding of quality** to the **freedom of work** and the **allocation and cycles of budgets**.

Consider the development side: for decades, these teams operated within a traditional **waterfall methodology**. Large-scale projects, stretching over years, progressed through rigid, sequential phases: analysis, design, build, test, and finally, deployment to production. Each phase was executed by highly specialized teams: functional analysts crafting perfect design documents, development teams handing over code to testers to find bugs, followed by security tests, with quality approval resting solely with testers. The final step, deployment, was often a fraught event, followed by yet more tests and, critically, a "handover" to the operations teams.

This ingrained mindset fostered a pervasive culture of **fragmented responsibility and externalized quality control**. Developers inherently trusted testers to validate quality, testers trusted release management, and release managers trusted operations.

The primary aim was not necessarily to deliver optimal software end-to-end, but rather to **secure approvals** from the next team in the chain. This led to a pervasive **mistrust** where "teams were controlling other teams." Such a fragmented, control-oriented approach is inherently antithetical to the speed and reliability demanded by modern digital services. To truly improve reliability, organizations must **overcome the mindset that quality relies on teams controlling each other**.

The multi-tiered support structure in traditional IT, comprising 1st, 2nd, and 3rd level support, acts as a series of "cushions." Each layer, while perhaps intended for specialization and filtering, effectively "wraps our people in absorbent cotton," ostensibly to insulate them from the raw, unfiltered "customer's reality." This insulation, however, comes at a profound cost:

- **Delayed Solutions:** "We delay with each layer the solution." Every time an incident is handed off from one level to the next, valuable time is lost, as each new person involved in a complex incident "must first understand the incident." This often leads to frustrating repetition for the customer, who is forced to explain their problem repeatedly.

- **Information Siloing and Loss:** As tickets are "passing... from one team to the next, we tend to lose information." This lost context must then be painstakingly "uncovered again," prolonging resolution times and increasing effort.

- **Shielded Accountability:** The most significant consequence is that "**the developers and the product owner are shielded from the customer**." They "have no direct contact with the problems and requests raised by the users." They are not the ones picking up the phone when a user complains about a crashing application or a frustrating new feature. Crucially, they "don't feel the pain their changes and features are causing when they do not work like they should." This detachment is a severe barrier to operational empathy and a direct feedback loop, preventing developers from instinctively building more reliable systems from the outset.

The irony is profound: in an era where customer experience is paramount and IT is the "heartbeat of enterprise success," the very people building that IT are insulated from its direct impact on the customer. This structure, while providing a semblance of order, inadvertently creates a system where the attention and incentives are misdirected, not focusing on the customer where it truly matters. This multi-layered, siloed structure inherent in traditional IT operations inherently creates significant drag, **slowing down the delivery of solutions to the customer**. When a problem emerges, the most logical and efficient approach would be to "first … ask the person who wrote the code." Yet, in the traditional model, the developer is often "asked last," after the incident has traversed multiple support levels. This delay, while perhaps perceived to offer benefits like "lower costs" or preventing "developers … getting distracted," comes at a profound cost to the customer experience and ultimately, the business. A strong focus on cost reduction, while understandable, is no longer the primary objective in a modern enterprise striving for speed and customer focus.

Beyond slowing down customer solutions, the traditional, fragmented structure paradoxically **increases the overall effort required from IT teams**. Each additional layer in the support hierarchy, and each handover point, inevitably "increases the communication effort." For every new feature deployed, a constant, often manual, flow of information is required from developers to all the various support teams. This constant "push" of information consumes valuable time and resources that could otherwise be spent on innovation or proactive reliability improvements.

The cumulative effect of communication overhead, information loss, and fragmented responsibility is a pervasive **"tension" within the IT organization**. This tension manifests in several ways:

- **Misunderstanding:** Despite efforts to communicate, handovers between different teams often lead to "misunderstanding," where the nuances of a problem or solution are lost in translation.

- **Mistrust:** When issues arise, especially after a handover, "mistrust" can quickly fester, leading to finger-pointing rather than collaborative problem-solving. "Who caused this problem?" becomes the implicit question, rather than "How can we fix the system?"

- **Rejection:** This tension can escalate to outright "rejection," particularly when operations teams, having recently dealt with issues from a previous change, are asked to deploy something new from development. The accumulated "pain" of handling problems "from the developers" creates a strained relationship.

This dynamic is palpable, as observed when talking to support personnel: they are "**worn down between the customer and the developers**," caught in a perpetual crossfire of complaints and blame. This unsustainable pressure leads to **high fluctuation** within operations teams, further increasing effort as new staff must be onboarded and ultimately **decreasing the quality of support**, pushing customers toward competitors. To reverse this cycle and "increase our customer intimacy, optimize time to market, and reduce the effort at the same time," the solution is clear: "**building build-run teams**." These teams, intrinsically responsible for both development and operations, fundamentally "change the structure of your teams, and with that, the structure of our organization."

Mindset and Entitlements: Overcoming the Deeply Ingrained "Fear of Failure" and Rigid Access Controls

The structural divide between development and operations teams is further exacerbated by a striking disparity in the **autonomy and flexibility** afforded to each group. Developers, especially those immersed in Agile methodologies for years, have often enjoyed considerable freedom in choosing their workplace and working hours. Their mantra: "Get the work done." Long before the pandemic, it was common to see developers working remotely, from a coffee shop, or even by the beach. Their schedules might tighten around releases to fix bugs rapidly, but fundamentally, they held the reins on their timelines, deciding when a feature was "ready to ship" and deploying to production when they felt "comfortable." Agile, with its emphasis on flexible timelines, amplified this sense of autonomy.

In stark contrast, operations teams typically experience the exact opposite reality. Their work is dictated by the relentless urgency of incidents. A ticket is assigned, and they are expected to "**start working immediately**." The clock is always ticking; a customer is waiting, potentially blocked from crucial business functions. "Ops teams have not much choice," observations confirm, often experiencing timed lunch breaks and a constant vigilance over their mobile devices for high-priority alerts. When a critical incident strikes, they have mere minutes to react, as the financial and reputational stakes are immediate and substantial. The halting of an assembly line due to an IT problem, as an analogy, can literally make front-page news.

Moreover, the very nature of their access to sensitive data—from account information to financial transactions and personal employee details—often confines operations teams to **"secure bay areas"** within the office, physically separated with locked doors, restricted access, and no personal devices. The challenges faced by these teams during the COVID-19 pandemic, adapting to remote work with such stringent physical and access controls, vividly illustrate this inherent rigidity.

Therefore, any meaningful attempt to improve reliability and transform the organizational structure must be acutely sensitive to these **divergent working models**, the **distinct triggers that initiate work**, and the fundamentally **different understandings of work itself** that have evolved over decades. Projects, for instance, once dictated development efforts, with clear beginnings and ends, sometimes even leading to the dissolution of teams once a product was "done" and handed over to "run" teams. Agile and DevOps, while revolutionary for development by breaking down silos between development, testing, and security through concepts like continuous integration (CI) and continuous deployment (CD), largely left operations untouched. Despite "Ops" being in its name, DevOps, in traditional companies, often stopped at the production deployment gate.

This has created a two-speed IT organization: one part continually adopting new skills, tools, and methodologies (design thinking, AI pair programming, test-driven development), while the other, the operations teams, often remains "untouched over years," still operating under principles akin to early ITIL frameworks. When initiating transformative change, it is crucial to recognize that one side of the house feels **overburdened** and constantly reactive, while the other feels **"left out"** of the innovation cycles, leading to deep-seated resentment and resistance.

This bifurcated reality—two different working models, driven by distinct incentives and performance metrics—has naturally cultivated **two fundamentally different belief systems** within the IT organization.

On one side, the **development teams** are primarily incentivized by the imperative to "get work done" and deliver new features to production quickly and efficiently. Their pressure often emanates directly from the business side, which demands speed and cost-efficiency. Developers strive for "clean code, well-structured and understandable, to make their work easy." Their internal compass points toward functionality, innovation, and rapid iteration.

On the other side, the **operations teams** are driven by a singular, overriding objective: **stability**. Their aspiration is "no interruptions. No weekend work, no late hours, because a high-priority ticket is still unsolved. Just a stable system." Their reality, however, is a constant barrage of user frustration, phone calls about non-functional systems, and the relentless pressure of critical tickets where "the clock is ticking" and the company is "losing money." Compounding this, Ops people are often subjected to meetings where leadership "searches for a scapegoat" to blame for system failures or poor application performance, such as long latency times. This relentless, often unappreciated, battle for stability breeds a deep-seated apprehension: **they know that "with each deployed change from the development team, their system gets more unstable."** This inherent tension, where one side's "done" signals the other's "starting gun" for firefighting, is a significant barrier to holistic reliability.

This enduring pressure and blame culture has deeply imprinted a **mindset driven by fear** within operations teams. In such an environment, every action—or inaction—can potentially lead to the next outage. Even the prospect of organizational change or shifts in team alignment is viewed through the lens of potential disruption and downtime. This fear of change translates into a pervasive **"yes, but..." attitude**. Operations teams, when presented with new initiatives, often respond with valid, yet ultimately paralyzing, caveats: "Yes, this is right, we must change this, *but* ... it will take a long time... we don't have the skills to do it ... I need approval from the business first ... we have to allocate a budget first."

This deeply rooted experience leads to an **inherent blocking of anything new**. The prevailing sentiment becomes, "All is working right now; we should not change anything." This resistance is not malicious; it's a learned survival mechanism in a system where change often correlates with pain. Over years, even decades, these two distinct "sides of the wall"—development and operations—have cultivated their own unique behaviors, habits, interpretations of situations, thinking patterns, and **protective reflexes**. Any perceived attempt to "slow down my work" or "make my application unstable" is met with resistance, stemming from years of deeply ingrained, often painful, experience.

Further dissecting the differing mindsets and experiences between development and operations, it's crucial to examine the contrasting **team structures** that have historically prevailed.

Development teams, especially those that have embraced agile practices, are typically organized into **small, cross-functional units**—often comprising 6–12 individuals, commonly referred to as "squads" or "pods." These squads, in turn, are grouped into larger "fleets." A hallmark of agile adoption is that the **management layer has actively "bought in" to decentralize authority**, empowering these squads to a significant degree. Their mandate is often framed around outcomes: leadership defines high-level epics, and the squads determine *how* to achieve them. In many customer-facing applications, IT leadership has even transferred authority over product direction directly to the business, leaving IT leadership primarily focused on technical debt, re-platforming, and application rationalization—in essence, "IT for IT" decisions. This necessitates **strong alignment between business and IT**, a partnership that ideally extends across all levels of the development organization, with product owners at the squad level and business owners at the management tier.

Conversely, **operations teams** often exhibit a vastly different, more traditional structure characterized by **large, hierarchical teams**. It's not uncommon to find 50, or even up to 100, people, reporting to a single manager. These teams are typically segmented into **1st, 2nd, and 3rd level support** as described in Chapter 2.

The skillset progression across these levels is stark: from basic phone support (1st level), to functional application understanding (2nd level), to full software engineering and troubleshooting under pressure (3rd level). This dictates application coverage: 1st level can support hundreds of applications, 2nd level a handful, and 3rd level typically focuses on one to three complex applications.

This hierarchical structure also shapes leadership perspectives. Managers of traditional operations teams often take pride in the sheer number of people under their direct control—a "dominator hierarchy" where quantity of reports signifies power. Phrases like "I'm responsible for 100 people" reflect a deeply ingrained mindset that views a large headcount as a personal achievement. When breaking down such silos and shifting toward smaller, cross-functional teams, it is imperative to address this deeply rooted mindset and the perceived "loss of power" early in the transformation.

Another significant impediment embedded within traditional organizations is the rigid control over **tools and working environments**, enforced by a labyrinthine system of "entitlements." This creates a stark division in the digital workspace that mirrors the organizational split between development and operations.

In highly regulated sectors—such as banking, insurance, defense, and pharmaceuticals—development teams typically operate within isolated environments: development, quality assurance, and user acceptance testing. Their daily work revolves around tools like Jira for issue tracking, Git for version control, and various unit and functional testing tools, all often integrated within their chosen Integrated Development Environments (IDEs). **Access control** is meticulously enforced, dictating who can view source code or access specific environments. Every change is rigorously tracked: who made it, when, why, and based on what request (e.g., from a product owner or due to a technical issue).

On the other side, **operational teams** primarily engage with **monitoring tools** and different **issue tracking systems**, predominantly dominated by platforms like ServiceNow or BMC Remedy. Their workflows are rigidly structured around ITIL processes, managing incidents, problem tickets, events, and change tickets. Some operations personnel possess direct access to **production databases**, administrative GUIs (graphical user interfaces), and live production applications and log files. They might even be authorized to directly modify production databases in emergencies—access deemed essential for minimizing customer interruption.

The critical issue is that **one team typically has no access to the tools used by the other side**. Developers cannot simply download or install Ops tools, and vice versa. Entitlements are binary: you are granted access for one domain or the other, rarely both. While developers might "tap into production" in an emergency, this is a highly regulated and supervised exception, not a standard operating procedure.

This raises a fundamental question for any organization embarking on an SRE transformation: **"How much time does it take to change the access control/entitlements in your company?"** Furthermore, is it truly feasible to "create a neutral role, not Dev, not Ops, but still save enough for your regulators?" The answers to these questions are complex and often expose the deeply ingrained bureaucratic hurdles that actively hinder the fluid, cross-functional collaboration central to SRE. Modifying these deeply embedded access controls is often a multi-month or even multi-year endeavor, entangled with compliance, security, and internal audit processes, making the integration of build-run teams an uphill battle. So the main question becomes: Is it worth it? Given the sheer scale of effort required—navigating layers of policy, satisfying rigorous audits, and reshaping long-standing security postures—organizations must weigh whether the promise of streamlined, resilient operations truly justifies the significant investment of time, resources, and organizational capital needed to overcome these entrenched barriers. We will explain and show here in this book that it is absolutely worth it.

The rigid organizational structure and tool segregation have, over decades, profoundly shaped the **career paths and skill sets of IT professionals**. In traditional enterprises, individuals tend to spend their entire careers firmly rooted on either the development or the operations side, rarely venturing across the divide. This specialization, while fostering deep expertise in one domain, creates a significant **skill gap** when attempting to merge these functions into a unified SRE or build-run team.

Career progression within large organizations often relies heavily on **established networks** and accumulated trust within a specific domain. Stepping outside of this established network, attempting to "change sides," often means starting over, building connections and trust from scratch. This inherent inertia is reinforced by incentive structures that reward specialization and discourage cross-domain exploration. Organizations, through their systems, inadvertently "dominate people" by not supporting the easy uptake of projects or tasks outside their defined roles. Applying for a new internal team often feels like an external job application, with the risk of rejection. The reality, as observed, is that **"fluctuation between Dev and Ops is rather low."** To be accepted in the other domain, one might even feel compelled to "hide your origin."

A pervasive, albeit often unspoken, **bias exists: "Developers are looking down on Ops people."** While this is a difficult truth to confront, it manifests in hiring practices and team dynamics. An Ops person applying for a development job faces significant hurdles; their skills in operational stability and incident response, while critical for SRE, are often undervalued by developers seeking pure coding prowess. Conversely, when a developer applies for an Ops role, there's a greater implicit trust that they "will figure it out," even if they lack immediate knowledge of monitoring systems. This inherent bias creates a structural disadvantage for ops professionals seeking to pivot into a more engineering-centric role, despite the fact that "software development is the most complex skill to learn."

Therefore, when designing an SRE transformation, it is imperative to acknowledge and proactively plan for this deeply ingrained "**bias**." Any approach must account for the reality that "most IT practitioners spend their whole career on one site or the other." It requires dedicated programs for **upskilling** (especially in software engineering for operations staff), **reskilling**, and **fostering a culture of mutual respect and learning** that transcends historical perceptions. Without addressing these deeply human and career-driven realities, the promise of build-run teams and true operational empathy will remain an elusive ideal.

The Weight of the Past: Confronting Complex Systems, Legacy Technologies, and Hidden Technical Debt

Beyond the human and organizational impediments, legacy enterprises confront a formidable technical challenge: their IT landscapes are characterized by **"highly complex IT systems"** that have evolved, sometimes haphazardly, over 10, 15, or even 20 years. This historical layering results in an intricate tapestry of technologies, often including relics like **mainframes** and esoteric languages such as **Smalltalk** (first released in 1972).

The narrative of this complexity is often one of **"missing architectural guidelines, missing technical blueprints, some budget stops, and fluctuation of people, with them the knowledge."** This erosion of institutional knowledge, coupled with the relentless pressure for new features and cost optimization, has inevitably led to a **"poisoned IT structure."**

Consider these illustrative anecdotes, drawn from real-world enterprise experiences:

- A leading automotive company, for instance, found itself in a precarious situation where it could **no longer locate the source code** for a critical application running in production. The original developer had departed, and with them, the institutional memory of that system had vanished. When a problem arises, there's no map, no guide, only an archaeological dig through digital ruins.

- In another instance, a team struggled to migrate a source code base to a modern version control system because the underlying framework was **12 years old**, with updates constantly postponed due to the "too much work" or "it's still running, why do we have to do it?" mentality. The inevitable outcome? The system eventually failed, necessitating a costly, time-consuming **rebuild of most of the software**.

This accumulation of **technical debt** is a pervasive and demoralizing problem. Teams often actively avoid engaging with old, poorly understood codebases, creating "new APIs around it" or "delaying handovers." In extreme cases, engineers even "leave projects because some source code is not understandable anymore." This creates a vicious cycle: the older and more complex a system becomes, the harder it is to maintain, leading to more technical debt and further eroding the talent pool willing to work on it.

The reality, then, is a landscape riddled with "source code, old frameworks, infrastructure, and antiquated versions." Any SRE transformation must confront this formidable legacy head-on, recognizing that it's not a mere technical hurdle but a deeply entrenched systemic challenge requiring **significant investment, strategic reprioritization, and a long-term commitment** to untangling this intricate web of technical liabilities. As Chapter 1 highlighted, traditional companies "born before the cloud" face unique challenges in implementing SRE, precisely because of this pervasive legacy infrastructure and mindset.

Agile, DevOps, Cloud: Where Operations Were Left Behind: Unpacking the Untouched Silos in ITIL-Driven Organizations

Unlike cloud-native startups, legacy enterprises carry a rich history of ongoing technological and organizational change. This history is characterized by successive waves of transformation, each attempting to address fundamental shifts in the IT and business landscape.

The first major wave saw teams transition from rigid **waterfall practices to agile thinking**. The core revelation was that a singular focus on cost reduction, prevalent in earlier eras, was no longer the "holy grail." Instead, the new paradigm emphasized **working in short iterations, with continuous validation from the client**. The Agile transformation's primary impact was to **break down the silo between business and IT**, centering all efforts around the customer. By integrating product owners into IT teams (pods or squads), a greater degree of independent decision-making became possible within the squad, extending even to portfolio and leadership levels through business and development partnerships.

Following this, **DevOps became mainstream**, focusing on automating the entire process from development to production deployment. This movement **broke the silo between development, quality teams, and security**. However, a critical observation from experience is that despite "Ops" being in its name, DevOps in traditional companies often stopped short of fully encompassing operations; production workflows largely remained "unthought."

The third significant movement has been **cloud-native adoption**. This shift primarily targeted **infrastructure teams**, addressing long-standing bottlenecks where developers often faced delays of days or weeks waiting for new environments or test data. Cloud computing, with its promise of agility and scalability, empowered infrastructure teams. Large cloud providers (Azure, AWS, Google) built services atop their infrastructure, enabling a "platform" approach. This led to the emergence of internal platform teams within enterprises, mimicking public cloud providers by building internal services and GUIs to automate manual work and provide "guardrails" to product teams, thereby reducing wait times. Infrastructure teams themselves had to adopt new skills, essentially becoming "squads" developing internal-facing applications for their internal clients.

The current situation in most legacy companies is therefore a **hybrid reality**. While the ambition is for development to be Agile and leverage DevOps, and for infrastructure teams to utilize public cloud providers and build internal platforms, the reality is a heterogeneous landscape. Some teams have indeed moved "really far" and fully adopted these new ways of working. Yet, "some or maybe many teams are still in waterfall mode, have not even automated unit testing, and deploy manually with some Ant scripts." This inherent **diversity and varying maturity levels** across different parts of the same organization are what make legacy companies "complex." Any SRE transformation must navigate this fragmented landscape, recognizing that a "one-size-fits-all" approach is unlikely to succeed.

A critical, yet often overlooked, consequence of these successive transformation waves (Agile, DevOps, Cloud) is that **operations have largely been left untouched**. Despite the prevailing narrative of being "customer-centric," the very teams with the most direct, intimate contact with the customer—operations and production support—have been excluded from this "customer-centric" optimization culture.

Operations teams in most traditional companies continue to function in **large, hierarchical structures**, often rigidly adhering to the principles enshrined in the "five books of ITIL." They remain the "blamed" party when an outage occurs, yet their invaluable "knowledge and intimacy with the customer" often goes unheard in the broader strategic discussions. While product development is informed by sophisticated A/B testing, design thinking workshops, and customer surveys, the raw, unfiltered voice of the customer complaining about a broken application often stops at the operations team, failing to reach the ears of those making product decisions.

This creates a peculiar paradox: a company striving for ultimate customer experience is effectively deafening itself to the direct feedback channels from its most affected users. The operational pain points, the frustrations experienced by the very people using the software, become isolated within the operations silo, preventing systemic learning and improvement where it matters most.

One of the most formidable barriers to the effective adoption of site reliability engineering is the enduring **divide between the traditional "development team" and the operations team**. In the Agile paradigm, a "development team" is envisioned as a cross-functional unit capable of handling everything from design, build, and test to deployment into production. However, this ideal often falls short when it comes to the crucial "business of operating the service in production."

Traditionally, the "application owner," typically associated with development, presides over *how* the product is built and *what* it does, maintaining close connections with the business. In contrast, the "service in production" is often owned by a separate "operations lead or ITSO," who remains "detached from the business." This disconnect means that many traditional conceptions of Agile either entirely overlook operations or address it only superficially. **SRE steps into this void, offering a comprehensive framework to bring operations firmly into the domain of the application owner, fostering holistic ownership and accountability**.

In the contemporary landscape of software development, most organizations, to varying degrees, have been shaped by the principles of **Agile methodology**. Agile has evolved from a niche practice into a mainstream approach for organizing work, spawning numerous team-level practices, scaled agile frameworks (like LeSS or SAFe), and systems for managing entire portfolios. It has fundamentally reshaped how work is organized and how budgeting decisions are made across all levels of an organization.

For teams embracing DevOps and SRE, Agile is often a prerequisite, almost a "foregone conclusion." Over time, the influence of agility has permeated areas far beyond mere work organization. Concepts like **Test-Driven Development (TDD)** (introduced in *Extreme Programming Explained* in 1999) and its evolution, **Behavior-Driven Development (BDD)**, illustrate how precise technical processes can be integrated into the Agile framework. At its purest, Agile organizes work around **product functionality that delivers value to the end-user**. Each work item should directly extend the service provided to the user. BDD, for instance, directly links application code to its functional value, often enabling test automation frameworks to connect human-readable acceptance criteria to specific parts of automated quality assurance, effectively annotating code with verifiable indicators of customer value.

This profound shift empowers the **application owner—the key decision-maker** in work organization and prioritization and the direct link to the business—to write specifications that are directly tied to code, fostering a much closer connection to product development and validation. The Product Owner, representing the application owner at the team level, has seen their role expand significantly, evolving from merely being the "voice of the customer" to an overarching "content authority" for all teamwork. As development teams have absorbed more disciplines (quality engineers, architects, DevOps engineers) into their cross-functional structure, product owners have had to develop a nuanced understanding of prioritizing a broader set of tasks—from CI/CD pipeline improvements to new functional features. While these disciplines might not always reside within the same team, in scaled Agile environments, they often draw from a shared backlog, governed by a prioritization process that dictates work allocation and budget distribution.

The challenge then becomes clear: how does one authority effectively make decisions that consider all these diverse domains? This is where SRE, much like BDD, provides a crucial "frame of understanding" to simplify this complex decision-making process.

The core principle guiding work prioritization remains the **value provided to the user**. To apply this principle to operational work, product owners need to understand two critical facets of the service: **reliability** (the level of service provided to the user) and **operability** (how easy the service is to operate). Armed with a clear grasp of these two measures, a product owner can confidently integrate operational work items into the backlog alongside user-facing functionality and other technical enablers. This integrated view allows for a holistic and data-driven approach to prioritizing work that truly optimizes for end-to-end customer value.

The concepts of reliability, stability, and service level are not new in IT operations. Organizations have long used service level agreements (SLAs) and other traditional metrics to gauge operational success. However, SRE introduces a fundamental shift: its measures of reliability must be **explicitly customer-centric**.

As will be discussed in depth in a later chapter (and as hinted at in Chapter 2, where SLOs were introduced as a "glue"), SRE meticulously defines target reliability by starting from the **user journey**. This means "explicitly pinning our service level indicators (SLIs) to the way the customer experiences the service." This creates a direct parallel to the Agile methodology's focus on delivering customer value: operational success is now directly linked to the customer's real-time experience, moving beyond mere technical uptime to encompass the true user perception of the service.

The "Optimal Reliability is a Business Decision" diagram (Figure from Chapter 6) powerfully illustrates this point: reliability, while crucial, also incurs cost. The goal is not infinite reliability but **optimal reliability**—the point of diminishing returns where additional investment in resilience no longer yields proportional business value. SRE provides the quantitative framework (SLIs, SLOs, Error Budgets) to have this critical conversation, ensuring that reliability is a conscious, data-driven business decision, not an abstract technical aspiration.

While reliability focuses on the *user's* experience, **operability** focuses on the *team's* experience—specifically, **how easy the service is to operate**. A system might be highly reliable from a user's perspective (e.g., rarely down), but if it requires immense manual effort, constant babysitting, or complex tribal knowledge to keep it running, its operability is low. This leads to burnout, high operational costs, and ultimately, compromises long-term reliability. SRE, through concepts like **toil reduction** (as discussed extensively in Chapter 6), directly addresses operability by transforming manual, repetitive tasks into automated, self-healing systems. A system that is easy to operate is intrinsically more sustainable and resilient.

For leaders responsible for ensuring their IT products enable continuous business operations and deliver value to customers, a holistic understanding of **all dimensions of stability** is paramount. This demands a fundamental shift in daily routine across every level of IT—team, management, and leadership—starting with a rapid, insightful glance at a **resiliency dashboard**. This dashboard should intuitively convey the current stability of IT products, with each team displaying individual resilience and performance measurements that indicate their proximity to predefined thresholds.

Imagine a scenario: it's a day of significant market volatility due to unforeseen global events, or perhaps a major industry shift. Customer behavior is changing rapidly, cybersecurity threats are escalating, and the stock market's reaction is unpredictable. In such a high-pressure environment, a leader needs to instantly know: **"how close it is to the next outage?"** Are systems still in the "green" (healthy) or already "amber" (under pressure)? Is it time to engage the crisis team and initiate immediate actions? If so, where: Europe or New York? Is the issue in the public cloud or on-prem infrastructure?

The power of an SRE-infused organization lies in its ability to provide this clarity. A leader opens their console and immediately sees that a particular service is "under pressure," its **error budget "nearly already consumed."** This granular, real-time information, aggregated from individual teams up to portfolio and domain levels, provides the actionable intelligence needed to make informed decisions.

At the team level, a software development team can, in real-time, see "how my product is working right now." They can compare current performance against historical data, identify systems under stress, and understand how customers are experiencing their products—whether latency is increasing or error messages are escalating.

This constant, pervasive observation empowers teams to "**immediately shift their focus to improve stability**." Developers can halt work on new features, pivoting instantly to production support, running diagnostics, analyzing system behavior, finding patterns, and driving root cause analysis to enhance reliability. This is the essence of **combined build-run activities**: teams are equipped and accustomed to switching immediately to production support, possessing both the knowledge of tools on both sides (development and operations) and the necessary entitlements to use them seamlessly.

Critically, this informed, surgical response means that while one product or service might be under stress and demanding focused reliability work, "other products are pushing new features at normal speed." This ability to act precisely and punctually, knowing which products are impacted, provides a distinct competitive advantage.

Over time, this continuous cycle of measuring resilience and balancing stability with feature velocity leads to systems that are "**getting more and more resilient overall**." This resilience extends not just to products but also to processes and people. Because resilience is systematically measured, teams avoid reactive panic, instead acting "punctual and well informed." Changes in market demand, or even outages from global cloud providers, no longer have the same catastrophic impact; products are designed to react autonomously, and teams are confident in their simulations. This shift reduces the human burden during crises, as much repetitive manual work is automated, and data is transparent and readily available, eliminating the need for "all hands on deck." The company progressively increases its resilience through a virtuous cycle of **observing, deciding, reacting, and learning**.

Beyond the technical and business benefits, this transformation also profoundly improves **employee satisfaction** and reduces fluctuation. By enabling smarter deployments (e.g., deploying when code is ready, not just on Fridays), the burden of weekend work and late shifts is significantly reduced. The inherent cross-functional nature of the teams creates more varied and engaging work, while increased decision-making autonomy empowers practitioners. This holistic approach ensures that SRE not only builds stronger systems but also fosters a healthier, more engaged workforce.

Architecting for Resilience: SRE's Transformative Impact on the Operating Model

Transforming an enterprise to achieve higher levels of resilience and stability doesn't necessitate building an entirely new IT operating model from scratch. Instead, it requires a strategic **update**, **change**, and **adjustment of the existing model**. This approach, deeply rooted in the SRE principles, guides the entire transformation journey. These guiding principles are formalized into "epics"—broad, directional statements that provide maximum transparency and flexibility for adaptation and recalibration, acknowledging that "each company is different; tailoring and adjusting is key."

This is a comprehensive transformation, touching upon all critical dimensions as described in Chapter 2. For instance, if the strategic goal is to "measure everything," the implications cascade across all four dimensions:

- **People:** Individuals must be upskilled to understand *how* to measure reliability and interpret the resulting data effectively.

- **Technology:** The necessary tooling for measurement must be implemented, allowing for data aggregation at team, domain, and leadership levels.

- **Structure:** Reporting and governance frameworks need to be updated to effectively utilize these new metrics, enabling appropriate action at the right organizational level.

- **Process:** Existing processes must be adjusted to empower individuals to take immediate action when established thresholds are exceeded.

The core SRE principles serve as the lodestar for this transformation. They are translated into actionable epics that drive specific changes within the operating model:

- **Measuring Everything:** The fundamental principle that enables data-driven decision-making, empowering individuals to act on facts and validate the impact of their work. This means moving beyond anecdotal evidence to concrete, quantifiable insights.

- **Leverage Tooling for Automation:** The strategic imperative to scale products without proportionally scaling the workforce. This involves automating repetitive tasks ("toil") and implementing self-service capabilities that empower teams to operate more efficiently.

- **Accept Failure as Norm:** A critical cultural shift that allows the enterprise to balance agility with resilience. By acknowledging that failures are inevitable, organizations can proactively plan for them, learn from them, and build systems that are inherently more robust. This also fosters a blameless culture, as discussed in Chapter 4.

- **Simulate and Test:** The practice of understanding system boundaries and pushing them further through controlled experimentation. This involves techniques like Chaos Engineering to uncover hidden weaknesses and learn how systems behave under stress. Each learning cycle informs further optimization.

These epics serve as the initial launchpad for the transformation, from which more granular "stories, tasks, and features" are derived to systematically adjust the existing operating model and integrate new capabilities. This iterative, adaptive approach is crucial for navigating the unique complexities of each legacy enterprise, ensuring that the SRE transformation is not a rigid imposition but a flexible, continuously evolving journey.

From Silos to Synergy: How SRE Principles Reshape Organizational Capabilities and Team Interactions

The concept of "You Build It, You Run It" is not merely an organizational philosophy; it is the **cornerstone of site reliability engineering** and a profound departure from traditional enterprise IT structures. This fundamental organizational split often extends vertically up the leadership hierarchy, with separate leadership teams for development and production support, influencing everything from **worker contracts and job descriptions to payment structures and sourcing/shoring strategies**.

This traditional separation also introduces complexities in financial accounting, particularly regarding **CapEx (Capital Expenditure)** and **OpEx (Operational Expenditure)**. As Chapter 1 touched upon, the disproportionate spending on operations versus engineering is a significant issue. Shifting to a build-run model can impact how costs are categorized and allocated, necessitating early engagement with the financial department to understand and plan for these implications. In essence, the "You Build It, You Run It" model aims to dismantle these long-standing silos, fostering holistic ownership and ultimately delivering more reliable and customer-centric products.

This division is not merely theoretical; it's manifested in distinct teams, reporting lines, and, as noted, even separate C-suite leadership roles. This organizational dichotomy, deeply entrenched over years, impacts every aspect of the employee lifecycle, from the very nature of their contracts and job descriptions to their compensation and the strategies for sourcing and shoring work. It creates an

environment where optimizing for cost efficiency in isolated silos often overrides the
holistic goals of speed and reliability, setting up an inherent conflict with the core
tenets of SRE.

Traditional sourcing and shoring strategies within large enterprises were
meticulously designed with a singular, overarching objective: **cost optimization**. This
deep-seated emphasis on cost efficiency, however, directly contradicts the multi-faceted
goals of SRE, which prioritize reliability, speed to market, and operational balance.

The "traditional thinking" behind these models involved **splitting activities and
geographically spreading them across different teams and locations**. This approach,
while effective at driving down labor costs, inherently **builds silos and adds layers of
checks and oversight**. Such fragmentation, ironically, is anathema to reducing time to
market and achieving the seamless flow of work required for modern, agile practices.
The deeply ingrained belief was that "business critical and complex work" should
remain "in-house and and more locally sourced and shored," while tasks deemed "not
critical to the business with lower complexity" were prime candidates for outsourcing.

This often led to scenarios where development work and operations work were
split between onshore and offshore locations. In the development process, for example,
"analysis and design work" might remain onshore and in-house, while "development
and testing" would be moved offshore to suppliers. Over time, this could even lead to the
paradoxical re-establishment of "specific testing capabilities again onshore, in-house," as
the limitations of fragmented quality control became apparent.

The complexity layers further when considering the **business criticality of
applications**. A financial services company, for instance, might completely outsource
the development and operation of an internal application for reserving parking spots,
deeming it "not business critical." However, the same company would ensure that a
"trading platform," considered "highly critical for their business," remains entirely in-
house and on-site.

Over decades, companies meticulously moved tasks around the globe, creating
a convoluted web of contracts and dependencies. While the latest strategic thinking
gravitates toward **value streams**, these are often constrained by existing, rigid
outsourcing contracts and a pervasive **lack of knowledge and skills** within the in-house
teams to simply "move back tasks."

This traditional structure, prioritizing cost reduction above all else, creates a direct conflict with SRE's goals. If the objective shifts to **reducing time to market, enhancing quality, and improving reliability**, then the optimal sourcing and shoring model looks vastly different:

- **Co-location and Unity:** "To have all people who are working on a product, for example, in the same room speeds up the communication." This colocation, whether physical or virtual, minimizes the communication overhead and inherent friction of handoffs.

- **Unified Company Teams:** "To have all from the same company also reduces overhead like approvals and multiple lines of command." This aligns incentives and accountability.

What is true for optimizing time to market is equally true for reliability and stability. Optimizing communication channels and fostering trust within teams directly "optimizes the system." This naturally leads to **in-sourcing projects and favoring nearshore and onshore models**—concepts that focus on **concentrating teams aligned to value streams at the same place, from the same company**. This profound shift in sourcing and shoring strategy is not merely a logistical change but a fundamental realignment of organizational priorities toward building inherent resilience and speed.

The multi-tiered support structure—1st, 2nd, and 3rd level—acts as a series of "**cushions**." Each layer, while perhaps intended to specialize and filter, effectively "**wraps our people in absorbent cotton,**" ostensibly to insulate them from the raw, unfiltered "customer's reality." The consequence: the urgent needs and direct feedback from the end-user are not immediately disturbing the teams who possess the power to change the application.

This layering introduces profound inefficiencies:

- **Delayed Solutions: "We delay with each layer the solution."** Each new person involved must first comprehend the incident, often leading to repeated explanations from the customer. "How many times have I been on the phone talking to the support, explaining my case again and again" is a common, frustrating user experience.

- **Information Loss:** As a ticket is "passed from one team to the next," crucial information is often "lost" or misinterpreted, necessitating "uncovering again." This leads to wasted effort and extended resolution times.

- **Shielded Accountability:** The most significant consequence is that **"the developers and the product owner are shielded from the customer."** They "have no direct contact with the problems and requests raised by the users." They are not the ones picking up the phone when a user complains about a crashing application or a frustrating new feature. They **"don't feel the pain their changes and features are causing when they do not work like they should."** This detachment is a severe barrier to operational empathy and a direct feedback loop, preventing developers from instinctively building more reliable systems from the outset.

The irony is profound: in an era where customer experience is paramount and IT is the "heartbeat of enterprise success," the very people building that IT are insulated from its direct impact on the customer. This structure, while providing a semblance of order, inadvertently creates a system where the attention and incentives are misdirected, not focusing on the customer where it truly matters. This multi-layered, siloed structure inherent in traditional IT operations inherently creates significant drag, **slowing down the delivery of solutions to the customer**. When a problem emerges, the most logical and efficient approach would be to "first… ask the person who wrote the code." Yet, in the traditional model, the developer is often "asked last," after the incident has traversed multiple support levels. This delay, while perhaps perceived to offer benefits like "lower costs" or preventing "developers… getting distracted," comes at a profound cost to the customer experience and ultimately, the business. A strong focus on cost reduction, while understandable, is no longer the primary objective in a modern enterprise striving for speed and customer focus.

Beyond slowing down customer solutions, the traditional, fragmented structure paradoxically **increases the overall effort required from IT teams**. Each additional layer in the support hierarchy, and each handover point, inevitably "increases the communication effort." For every new feature deployed, a constant, often manual, flow of information is required from developers to all the various support teams. This constant "push" of information consumes valuable time and resources that could otherwise be spent on innovation or proactive reliability improvements.

The cumulative effect of communication overhead, information loss, and fragmented responsibility is a pervasive **"tension" within the IT organization**. This tension manifests in several ways:

- **Misunderstanding:** Despite efforts to communicate, handovers
  between different teams often lead to "misunderstanding," where the
  nuances of a problem or solution are lost in translation.

- **Mistrust:** When issues arise, especially after a handover, "mistrust"
  can quickly fester, leading to finger-pointing rather than collaborative
  problem-solving. "Who caused this problem?" becomes the implicit
  question, rather than "How can we fix the system?"

- **Rejection:** This tension can escalate to outright "rejection,"
  particularly when operations teams, having recently dealt with issues
  from a previous change, are asked to deploy something new from
  development. The accumulated "pain" of handling problems "from
  the developers" creates a strained relationship.

This dynamic is palpable, as observed when talking to support personnel: they
are **"worn down between the customer and the developers,"** caught in a perpetual
crossfire of complaints and blame. This unsustainable pressure leads to **high fluctuation**
within operations teams, further increasing effort as new staff must be onboarded and
ultimately **decreasing the quality of support**, pushing customers toward competitors.
To reverse this cycle and "increase our customer intimacy, optimize time to market,
and reduce the effort at the same time," the solution is clear: **"building build-run
teams."** These teams, intrinsically responsible for both development and operations,
fundamentally **"change the structure of your teams, and with that, the structure of
our organization."**

Redefining Leadership: Aligning Dev and Ops Objectives for Shared Accountability and Value Creation

For any significant organizational transformation, particularly one as disruptive as the
shift to SRE and build-run teams, **leadership buy-in and continuous education are
paramount**. In traditional companies, leadership often operates within two distinct,
often conflicting, silos: "Production leadership fights against each unplanned downtime
with managing incidents," while "Project leadership is incentivized to deliver in time
with the least possible costs." These divergent experiences and values create an inherent
tension.

Therefore, a critical step is to **"educate our leadership together to share our new
ways of working and mindset."** This involves conducting joint workshops to align on
a new, unified set of objectives that **"are not preferencing stability or lead time for
change."** Instead, both stability and speed are recognized as equally important, to be
"driven from the same leadership together." To facilitate this understanding, concepts
central to SRE, such as the **Error Budget**, and the transparency offered by **SLIs and SLOs**
in making "the customer experience transparent," are introduced and explained. This
understanding is key for subsequent strategic decisions.

Crucially, leadership needs to clarify **"who has to sign off for the SLA."** This involves
explicitly calling out **"the role of the business leaders and the product owner in the
team."** This step helps to **"drive the change that involves a closer relationship to the
business department and cooperation based on the transparency to our customer
with SRE, SLI, SLO, SLA, and the Error Budget."** By clearly defining accountability at
the business level for service reliability, SRE becomes a strategic business conversation,
not just a technical one.

Furthermore, **showcasing early successes through pilot programs** is vital for
building momentum and convincing skeptical stakeholders. As discussed in Chapter 2, a
"pilot and learn" approach is recommended to validate the model in a real-world setting.
This provides tangible evidence of SRE's value, transforming abstract concepts into
demonstrable improvements. Leadership needs to see concrete results, even on a small
scale, to justify broader investment and cultural shifts.

The "People Follow Function" Imperative: Strategies for Creating Integrated Build-Run Teams

The core of optimizing for SRE People Follow Funtion in a legacy enterprise lies in the
creation of new build-run teams—teams that inherently bear responsibility for both
development and operations. This is a fundamental shift, and as Chapters 1 and 2
emphasized, it is where SRE truly demonstrates its strength by enabling teams to balance
reliability with change. However, traditional companies often harbor a bias against
build-run models, dismissing them as approaches suitable "just something for startups."
To overcome this deeply ingrained skepticism, the most effective strategy is to **"start
small and build a convincing showcase"** through pilot programs, demonstrating the
tangible benefits rather than relying solely on theoretical arguments.

The strategic selection of these initial pilot teams is paramount for success, as further detailed in Chapter 6:

- **Multiple Pilots:** It is advisable to "pick 2-3 teams" rather than relying on a single pilot. This diversification mitigates risks, as "too much can happen" with a single team (e.g., sidetracking, resistance, unexpected blockers).

- **Visibility:** The chosen product or application should be "well known in the company"—not necessarily the most critical, but visible enough that "the change must be heard and noticed." This ensures that successful outcomes generate internal momentum and demonstrate value across the organization.

- **Risk Mitigation: "Don't choose a high business critical application"** for the initial pilot. This provides "enough wiggle room and flexibility" to experiment, learn, and iterate without the immense pressure of immediate, catastrophic business impact.

- **Technological Familiarity:** Prioritize teams working with "common technologies," such as Java or .Net applications, as these often have broader skill pools and fewer unique legacy complexities than niche systems.

- **Team Willingness:** Most importantly, the pilot teams must exhibit genuine **"willingness"** and enthusiasm. If there's a need to "convince someone," it's likely "the wrong pilot team." This voluntary participation fosters a more positive environment for learning and adaptation.

This "start small" approach, focusing on tangible results and internal advocacy, is critical for building the momentum and evidence needed to scale the build-run model across the broader enterprise, systematically dismantling the "bias" against it.

The most effective, and perhaps most direct, way to create a build-run team with immediate impact is by adopting the principle of **"People follow function."** This premise is deceptively simple yet profoundly powerful for accelerating success in a transformation.

What it means in practice is this: when a function (a specific task or responsibility) is transferred from one part of the organization to another team, the individuals currently performing that function are also moved into the new team. So, if the objective is for a build-run squad to take on operational responsibilities, then the existing "operations experts" are directly moved "into the squad."

This approach offers several immediate benefits:

- **Accelerated Integration:** It cuts through the traditional hiring and training cycles that would otherwise be needed to onboard new team members with operational skills into development squads, or vice versa. Teams can "start immediately" with the necessary expertise.

- **Knowledge Transfer:** The deep, often tacit knowledge held by operations experts about the live production environment, its quirks, and its common issues is immediately integrated into the development team. This "operational empathy," as discussed in Chapter 1, becomes intrinsic to the build process.

- **Reduced Pain Points:** While this might seem disruptive, it bypasses many of the challenges associated with trying to teach new operational skills to developers from scratch or waiting for new hires. It shifts the focus to integrating existing expertise.

Once the initial team is formed by following the function, secondary steps can then be addressed, such as optimizing sourcing and shoring strategies and initiating cross-skilling programs among team members.

Subsequently, the focus shifts to establishing the "new working model, tooling, entitlements, and skills." This involves

- **Rotation Models:** Implementing rotation models between on-call duties and core development work, ensuring all team members gain practical experience in both realms.

- **Backlog Management:** Addressing the practical reality of different backlogs (e.g., operational issues in ServiceNow, development features in Jira) and deciding whether to combine them or maintain separate, but aligned, tracking systems.

- **Entitlement Alignment:** Ensuring that practitioners have the **"right entitlements"** to **"move seamlessly between operations and development,"** using the necessary tools in both domains. This requires a significant effort in updating access controls, as previously noted.

- **Usability and Clarity:** Recognizing that team members will now use "much more tools than in the past," it is crucial to ensure that **"usability is key," "processes must be straightforward,"** and **"priorities clear."**

By strategically moving people with their functions, organizations can rapidly establish the integrated teams necessary for SRE, accelerating the cultural and operational shifts that underpin resilient systems.

To be successful, we must overcome four main impediments when transitioning to build-run teams:

1. **Worker Contracts, Worker Councils:** The journey must begin with a critical and often sensitive discussion with the **Human Resources (HR) department**. This is because the shift fundamentally alters existing "worker contracts" and job descriptions. Traditional contracts meticulously define a person's role and link it to their salary. Introducing new tasks, such as on-call support, that may carry different demands (e.g., working outside standard hours) can impact compensation and require formal contractual adjustments. Furthermore, in many countries, particularly in Europe (e.g., Germany or England), significant changes to employee roles or working conditions necessitate the **acceptance and buy-in from worker councils or labor unions**. These powerful bodies protect employee rights and can become a significant impediment if not engaged early and transparently. Therefore, a foundational first step is to **engage HR and Worker Councils Early** to identify and address potential legal or contractual implications, allowing for the development of new role definitions (e.g., "SRE Engineer" or "Full-Stack Engineer with Operational Responsibilities") that are legally compliant and

acceptable. HR can also help **understand Legacy Constraints**,
identifying existing "legacy contracts" that might "hold us back."
This initial, often bureaucratic, step is crucial, as it **"determines
our approach and the timeline"** for the entire transformation.

2. **Developers Don't Want to Do Support:** One of the most
   persistent and emotionally charged objections to the "You Build
   It, You Run It" model from developers is the prospect of **"No
   support!"**, "I have not signed a contract to solve incidents," or
   "I don't want to work on the weekend or late at night." These
   reflect deeply ingrained expectations and personal realities, as
   developers often see themselves as "the creator of new features
   and fancy algorithms," viewing support work as less glamorous
   and inherently more stressful. While this "shipping problems
   to somewhere/to someone else and not wanting to deal with
   the consequences" is indeed a "red flag" signaling potentially
   "toxic behavior" and a lack of desired "collaboration" where "we
   win together, we lose together," a nuanced approach is critical.
   The core challenge is to **balance the imperative of end-to-end
   accountability with the diverse life scenarios and legitimate
   concerns of the workforce.** For example, a parent with young
   children might genuinely be unable to fix a ticket between 2 and
   6 PM, or someone new to operational responsibilities might
   require a "fall-back option when I have questions" during on-
   call shifts. The expectation of "overburdening working hours"
   is also unsustainable and unhealthy, leading to burnout and
   high fluctuation. To address these concerns while fostering
   accountability, a multi-pronged strategy is essential:

   - **Optimizing Operational Burden:** Proactively work to reduce
     the *need* for late-night or weekend work through deployments
     on Mondays, smaller/frequent deployments (reducing "blast
     radius"), "deployments whenever the code is ready," better/
     automated risk assessments, and automated testing.

- **Flexible On-Call Models:** Implement "shift models with rotation" or "follow-the-sun" approaches, though "follow the sun" needs calibration for low-severity issues due to its high cost (3 teams, each covering 8 hours, with at least three people per team).

- **Acknowledging Life Plans:** "We must establish a model for different life plans." While the team is ultimately responsible, individual contributions can vary, requiring empathy and support.

- **Prioritizing Prevention:** Most importantly, the core SRE strategy is to **"create a system and workflow to prevent tickets"** and "to see errors before a ticket gets opened." This proactive approach reduces the very incidents that lead to undesirable on-call burdens. By proactively addressing concerns and providing solutions that reduce pain, organizations can mitigate resistance and successfully transition to a build-run model with genuine end-to-end accountability.

3. **Changing Teams and Objectives for the Leadership:** This impediment is addressed by the need to **educate leadership** and align them on a unified set of objectives that balance both stability and speed of change, driven by shared leadership.

4. **Missing Skills from the Workforce:** The rigid organizational structure and tool segregation have profoundly shaped the **career paths and skill sets of IT professionals**. Individuals often spend their entire careers rooted in either development or operations, rarely venturing across, creating significant **skill gaps** when attempting to merge functions into SRE or build-run teams. This inertia is reinforced by incentive structures discouraging cross-domain exploration. Organizations inadvertently "dominate people" by not supporting easy uptake of tasks outside defined roles. The reality is **"fluctuation between Dev and Ops is rather low."** A pervasive, unspoken **bias exists: "Developers are looking down on Ops people,"** making it harder for Ops to

pivot to engineering roles despite software development being
the most complex skill to learn. Therefore, SRE transformation
must acknowledge and proactively plan for this "bias" through
dedicated programs for **upskilling** (especially in software
engineering for operations staff), **reskilling**, and **fostering a
culture of mutual respect and learning** that transcends historical
perceptions. Without addressing these deeply human and
career-driven realities, the promise of build-run teams and true
operational empathy will remain an elusive ideal.

The adoption of build-run squads fundamentally necessitates a re-evaluation and
reshaping of an organization's **sourcing and shoring strategies.** Sourcing dictates
whether work is performed "in-house or not," while shoring determines "on what
locations the team members are sitting." Traditionally, operations work has often been
"bundled in big clusters and outsourced more heavily than development work." Certain
quality assurance tasks might be kept in-house, while others are outsourced. Similarly,
"innovative, business-critical work" tends to be onshore, while "commodity work" is
more likely offshore. This cost-centric historical approach directly conflicts with the SRE
imperative to optimize for speed to market and reliability.

When optimizing for speed and reliability, the core objective is to **"enable and
empower people to act quickly when an outage happens."** This means eliminating the
need to "call someone to get permission to improve the stability of an application" or
navigate contracts with differing incentives. The driving force must be the **"experience
of our customers, not the lines of a contract."** Every "barrier, each kilometer, and
each line of paper on a contract is adding complications" to communication and
responsiveness. Each complication can translate into a "10–20% additional effort"
in communication. The ideal, "fastest and most reliable build-run team sits in one
room and has the same objectives." While the pandemic taught us compromises are
sometimes necessary (e.g., remote work), the underlying principle is to minimize
complications wherever possible. The consequence of this reprioritization is a
noticeable trend of **insourcing tasks "close to the customer."** This strategic shift aims
to achieve faster communication, tighter feedback loops, and a more unified approach
to reliability. By consolidating expertise and accountability within integrated teams,
organizations can more effectively respond to dynamic market demands and ensure the
resilience of their critical services.

Modernizing Operational Processes: SRE's Blueprint for Efficiency and Stability

Incident and problem management are the crucible of reliability in any IT organization. In traditional enterprises, these processes are deeply entrenched, characterized by multi-tiered support structures and a reactive posture. SRE fundamentally re-engineers this landscape, transforming it from a cost-efficient but slow, blame-prone system into a dynamic, data-driven engine for continuous improvement. This section dissects the traditional approach and presents SRE's transformative blueprint.

Beyond Reactive Fixing: The Shift to Proactive Incident and Problem Management: Moving from After-the-Fact to Anticipatory Action

As previously established, traditional companies have a tiered support structure consisting of **1st, 2nd, and 3rd level support**. This hierarchical layering, while seemingly logical for scale and specialization, inadvertently creates significant barriers to rapid problem resolution and holistic learning. As described, each of these "layers... is a cushion," designed to "wrap our people in absorbent cotton," ostensibly to protect them from the immediate "customer's reality." However, this insulation comes at a profound cost:

- **Delayed Solutions: "We delay with each layer the solution."** Every time an incident is handed off from one level to the next, time is lost. Each new person involved in a complex incident **"must first understand the incident."** This often leads to frustrating repetition for the customer, who is forced to explain their problem repeatedly.

- **Information Siloing and Loss:** As tickets are **"passing... from one team to the next, we tend to lose information."** This lost context must then be painfully **"uncovered again,"** prolonging resolution times and increasing effort.

- **Fragmented Accountability:** The separation creates a **"not my problem"** mentality, where each level's incentive is to hand off the issue rather than assume end-to-end ownership. As Chapter 2 highlighted, this leads to a **"fragmentation of ownership and visibility"** where **"no one individual or team possesses full end-to-end visibility or ownership of reliability."** This "hot potato" effect further delays resolution and erodes trust.

In essence, while traditional incident management aims for order through hierarchy, it inadvertently fosters a slow, inefficient, and often frustrating experience for both the customer and the teams involved. It's a system optimized for process adherence over rapid problem-solving and for cost efficiency (by leveraging lower-skilled resources at the front line) over customer delight.

A critical flaw in the traditional IT operating model is the inherent **lack of authority and influence that traditional operations teams wield over the reliability of applications**. Their primary focus is overwhelmingly **reactive**, centered on "supporting after the incident happened." While 3rd level support teams do possess access to source code and can implement changes, these are predominantly "bug fixes." Their work is driven by immediate "user problems," requiring "short notice" fixes to restore service.

Requests for long-term, systemic solutions, often captured in "problem tickets," are a different matter. In many traditional setups, these problem tickets linger, sometimes for "2–3 years," because "the team only works on problem tickets when they have enough time." The incentive structure is skewed: it's often "easier to use the workaround, like restarting a server, than really fixing the problem." Teams are **"incentivized to provide a fast solution"**—a quick fix—rather than a robust, lasting one. This operational model is thus **"optimized for cost"** and immediate service restoration, but **"not optimized for the customer experience relative to the costs"** of recurring issues.

This limited influence means that operations teams are perpetually caught in a cycle of firefighting without the power to proactively **"optimize the reliability of an application"** from its design or architectural roots. They are the symptom-fixers, not the root-cause engineers.

The multi-layered structure of traditional IT operations effectively creates a **protective shield around development teams and product owners**, insulating them from direct customer contact and the immediate consequences of their code. The very people tasked with making decisions on new features and writing the code—the "practitioners who are writing the new features and changing the applications"—have the **"least contact to the customer, to the person using the application."**

112

This detachment has profound implications:

- **Lack of Empathy:** Developers and Product Owners **"don't know the complaints," "they don't know if a new feature is used or not,"** and critically, **"they don't feel the pain their changes and features are causing when they do not work like they should."** This absence of "operational empathy," a concept highlighted in Chapter 1, severs the vital feedback loop between creation and consumption.

- **Misdirected Attention and Budget:** Despite the fact that **"most of the money is spent in production support"** (often "2–4 times as much" as on development, especially for legacy applications), product owners and businesses remain primarily **"focused on building new features."** They often only engage with the production support team **"when the application stands still,"** indicating that their attention is **"going in the wrong direction."**

- **The "Back Seat" Problem:** The metaphor is apt: **"Most Product Owners have found over the years a nice seat in the back of the car. Giving directions without seeing the street and the traffic."** They are disconnected from the immediate operational realities and the challenges of maintaining stability.

This systemic shielding prevents a holistic view of the product lifecycle, prioritizing feature delivery over the equally critical aspects of reliability and operational cost. It is a fundamental barrier to achieving the customer-centricity that modern digital businesses demand.

The very nature of traditional incident management is fundamentally **reactive**: it allows us to **"only see what happens after the fact."** When an incident ticket is opened, the problem—and its impact—has already manifested. This inherent "after-the-fact" focus fundamentally limits an organization's ability to be proactive and prevent disruptions.

While analyzing incidents through "problem tickets" *does* offer a mechanism to **"learn how we can optimize the application,"** it is still a form of learning from past failures. In the context of the **"complex systems"** prevalent in legacy companies, where **"not a single person understands the system anymore,"** relying solely on post-incident analysis for learning is insufficient.

To truly manage complexity and drive reliability, the focus must shift toward prevention: **"we must start learning how our system is working and take more symptoms into account than only incidents."** SRE, as highlighted in Chapter 1, aims to provide a "sixth sense, like a spider sense," to **"see before something happens."** The goal is to detect and address issues *before* they escalate to an incident, *before* the user experiences a problem, and *before* an incident ticket needs to be opened.

Historically, production teams *have* attempted to be more preventive, but constant **"pressure to reduce costs"** often forced them back into a reactive mode, prioritizing "solving incidents" over proactive improvements. This dynamic has created a continuous cycle of firefighting, perpetuating instability rather than building enduring resilience.

The cumulative effect of traditional IT structures—fragmented responsibilities, multi-layered support, shielded developers, and reactive incident management—results in a process that is **"highly cost efficient but slow."** This seemingly paradoxical outcome arises because each individual silo, optimized for its narrow function (often cost reduction), inadvertently creates friction and delay across the end-to-end value delivery chain.

As Chapter 1 eloquently put it, "IT processes are speeding up more and more but lack safety guidelines." This traditional setup, while making "sense in silos," operates on a logic of **"hand-overs from one team to another"** and a rigid adherence to "clearly described entrance and exit points." This creates a culture where "everybody protects its silo with borders," leading to a pervasive bureaucracy that **"slows us down."** For instance:

- "3rd level only accepts tickets when all required information is captured."

- "Operations only accept changes in production if the knowledge transfer was executed."

While these individual checks might seem reasonable from a control perspective, in aggregate, they impede flow. This siloed structure is diametrically opposed to the needs of a "small build-run team," where such fragmentation would cause the team to **"fall apart when the first team member wants to take vacations."** In build-run teams, the inherent interdependence means "they are in this together," necessitating a rapid internal shift toward collaboration and shared responsibility. The traditional model, therefore, represents a fundamental impedance mismatch with the modern enterprise's demand for speed, agility, and a holistic focus on customer value.

The Observe-React-Solve-Learn Cycle: Implementing a Dynamic, Holistic Approach to Production Support

For build-run teams to be effective, incident management cannot remain an isolated, reactive function; it must be **seamlessly integrated into the "normal workflow of application teams."** The challenge is to achieve this integration without constantly **"distracting"** the teams every time an incident arises.

The core principle here is that **"everybody is responsible for the stability of the product,"** while also being responsible **"to improve the product with new features."** This dual accountability requires a dynamic, data-driven structure within the team— one that guides its focus toward either stability or new features based on objective facts. This structure must be "visible for all members and formally validated by stakeholders." It's about proactive prioritization and resource allocation, rather than constant reactive firefighting.

This shift moves incident management from a specialized, isolated task to an inherent part of the application team's daily rhythm, ensuring that reliability is an ongoing concern, not just an afterthought.

To successfully integrate incident management into build-run teams and transition from a reactive, siloed approach to a proactive, holistic one, a three-pronged strategy is implemented:

1. **Team Rotation Model:** This is a crucial structural innovation. In a build-run squad (e.g., 10 people), a rotating subgroup (typically 2–3 members) is explicitly assigned **"responsibility for production support"** for a defined period (e.g., 1 week). This **"rotation model has multiple benefits"**:

   - **Dedicated Focus:** It **"gives developers the time to focus on building new features without interruption"** when they are not on rotation.

   - **Shared Responsibility and Empathy:** **"All squad members have the same responsibility and share the pain"** of operational issues. This fosters **"operational empathy,"** ensuring developers directly experience the impact of their code.

- **Immediate Start:** It allows for an immediate implementation of build-run, even if "strict entitlements" initially restrict access to production for all developers. Over time, these entitlements are broadened as trust and skills grow.

- **Knowledge Democratization:** It ensures that operational knowledge and troubleshooting skills are diffused across the entire team, rather than being concentrated in a few individuals.

2. **Observe, React, Solve, Learn Method:** This conceptual framework provides a lightweight, agile alternative to the often "clunky and too heavy" ITIL fundamentals for small, full-stack teams. It's a dynamic cycle designed for rapid, continuous improvement:

- **Observe and Anticipate (Proactive):** The team continuously monitors the application (both externally for known issues and internally for the unknown, i.e., "Monitoring" versus "Observability"), simulates problems (Chaos Engineering), and analyzes historical incidents to identify patterns. The goal is to **"mitigate the risk upfront"** and work on **"optimization items from the backlog"** to increase resilience and reduce toil.

- **Orient and Decide (Situational Awareness):** When an issue emerges (either via an incident ticket or proactive detection), an **Incident Commander** (often the Scrum Master in agile squads) **"orients and decides about the next steps."** This involves gathering context and **"swarming to inform and to get the right experts."**

- **React (Action and Resolution):** An **Operation Lead** (who can also be the Incident Commander in less severe cases) focuses on solving the issue, deploying fixes through the standard DevOps pipeline without shortcuts.

- **Learn and Optimize (Continuous Improvement):** After resolution, the team conducts a **blameless post-mortem meeting** to **"optimize the ROI for our company"** from the "unplanned investment" of the incident. Actions are drafted, the backlog is updated, and learnings are made transparent for the broader organization. This step reinforces that **"incidents are opportunities to harness the power of DevOps and SRE."**

3. **Incident Command System (ICS) Roles:** To manage incidents effectively, especially during "wild spreading and extending issues," a flexible role structure is adopted. While one person can hold multiple roles in normal times, in a crisis, dedicated roles ensure clarity and efficiency:

   - **Incident Commander (IC):** The overall **"person in charge"** of production support during their shift. They oversee ticket status, ensure operations leads accept tickets, validate severity, and drive escalations. The IC makes critical decisions when the business is impacted or threatened.

   - **Operation Lead (OL):** The primary person solving the incident, requesting support when needed, and focusing on resolution during the "Orient & Decide" and "React" phases.

   - **Communication Lead (CL):** Informs "stakeholders, leadership, and customers if required." While the OL might handle this in normal scenarios, a dedicated CL is crucial during major outages to ensure **"well-written and understandable"** communication at the **"right frequency"** to different audiences.

This structured yet flexible approach, supported by a "Manager on Duty" calendar and real-time dashboards, transforms incident management from a chaotic firefighting exercise into a disciplined, learning-driven process. The emphasis on **preventing "Ticket Ping Pong"** by using parent/child tickets, distinguishing between **Incident and Problem Management**, and focusing on **"Service Level" metrics** (solution time, reopened tickets), all contribute to a more effective and customer-centric operational posture. The ultimate shift is from merely solving problems *after the fact* to proactively managing risk and learning from every disruption.

The traditional separation of problem management from core development backlogs is a significant contributor to lingering technical debt and recurring incidents. In legacy enterprises, the production support team typically "opens the problem tickets," which are then handed over to the developer team for solution implementation. This "handover and the different objectives of the teams correlate to the disjunction of the process," often resulting in the "solution gets delayed and delayed more and more," with the "victim is the customer." Problem tickets, as noted earlier, can languish for years.

When organizations transition to **build-run teams**, this systemic issue is **"solved immediately."** At its core, the **problem ticket becomes "just another item in the developer backlog."** Critically, it is then **"prioritized from the same people who are feeling the pain in production."** This direct connection between operational pain and development prioritization eliminates the disjunction and formal bureaucracy of handovers.

The **Error Budget** (a concept extensively detailed in Chapter 2 and Chapter 6) plays a pivotal role in this integrated prioritization. It provides a quantifiable, objective mechanism to **"validate the priority between new features, toil reduction, and reducing technical depth."** If a service's error budget is being consumed rapidly due to recurring problems (tracked via problem tickets), the team has a clear, data-driven mandate to pause new feature development and prioritize work that improves stability, reduces toil, or addresses technical debt. This ensures that reliability is not an afterthought but a conscious business decision, driven by measurable customer impact.

Traditional ITIL frameworks advocate for "continuous improvement" (CI), where identified tasks for enhancement are added to a backlog. However, the unfortunate reality in many legacy organizations is that these CI items often **"stay in the backlog, and no one is working on them for quite some time,"** sometimes **"older than a year,"** and ultimately are **"closed not because the improvement was executed; teams were closing items because they were too old."** The core issue here is a lack of prioritization and a disconnect from the immediate operational pain.

SRE fundamentally elevates this concept from mere "continuous improvement" to **"continuous automation."** The shift is profound: when a build-run team identifies an opportunity for improvement, the immediate question is no longer just "can we improve this?" but rather, **"can we automate that?"** This **"automation built in the DNA of SRE"** ensures that improvements are not just theoretical or manual fixes but scalable, repeatable, and enduring solutions.

This focus on automation is driven by two key aspects:

- **Clear Prioritization with Error Budget:** As discussed, the Error Budget provides a direct, quantifiable signal for when to **"focus on improving the stability when required."** If the budget is burning, automation of toil or problem resolution becomes a high-priority investment to replenish it.

- **Focus on Scalability and Toil Reduction:** SRE recognizes that
  manual work ("toil") does not scale with system growth. By
  dedicating time to automate this toil, teams free up valuable
  engineering time for higher-impact, strategic work. As detailed in
  Chapter 6 ("Toil Management"), this frees up human capital and
  reduces costs.

- **Empowering Operations with Automation Skills:** Integrating
  automation skills directly into build-run teams ensures that SRE
  practitioners **"are not just improving; they improve by writing an
  automation script."** This transforms reactive operational work into
  proactive engineering, moving away from manual drudgery.

This systematic emphasis on automation, underpinned by the Error Budget
for prioritization, ensures that every improvement effort contributes directly to the
scalability and resilience of the system, fundamentally changing how IT operations are
perceived and executed.

The question of whether the traditional IT Helpdesk or 1st level support (often
synonymous with the initial "on-call" function) remains necessary in a fully SRE-
transformed enterprise is a critical point of discussion. These roles typically involve
individuals "picking up the phone when the customer is calling," "accepting the ticket when
the customer opens one," or "creating a new ticket when the customer writes an eMail."
While this service traditionally provides "decent quality" and "professional" customer
interactions, there are **"two schools of thought about On-Call"** in the context of SRE:

- **Helpdesk as an Unnecessary Silo:** This perspective views the
  Helpdesk as just **"another not required silo, shielding the squad
  from the customer."** It argues that direct customer contact is
  essential for operational empathy and rapid feedback, which the
  traditional Helpdesk layer impedes.

- **Helpdesk as a Quality Enhancer:** This perspective argues that
  the Helpdesk **"increases the quality in customer interactions,"**
  particularly through **"good language skills,"** and **"reduces
  unrequited interruptions for developers."** It recognizes that
  expecting developers to handle all customer interactions directly,
  especially those requiring specialized language skills for a global
  customer base, might be impractical.

Sure, we lean toward pragmatism: While direct customer feedback is invaluable, it's not always feasible or efficient for every developer to handle every call, especially across diverse languages. For interactions that do not require picking up the phone (e.g., customers opening tickets via a form), the build-run squad can leverage **translation services** to access unfiltered customer feedback.

However, the necessity for **on-call service with shifts** remains, particularly for global applications requiring 24/7 support. This can be a challenge, requiring careful planning of "shift models with rotation" or "follow-the-sun approaches" to cover different time zones, as discussed previously.

A crucial innovation in SRE is in **alerting mechanisms**. Moving beyond simple text messages (which are unreliable for time-sensitive alerts, as evidenced by delayed "Happy New Year!" messages), SRE advocates for sophisticated systems where **"a system calls our support and a robot voice reads the text from the system alert."** This ensures critical alerts are delivered reliably and immediately.

Ultimately, while the form and function of the traditional Helpdesk might evolve, the need for an initial point of contact and sophisticated alerting remains. The goal is to ensure that incidents are triaged efficiently, communicated effectively, and routed to the build-run teams with the necessary context, without creating unnecessary layers of delay or insulation.

Reimagining Segregation of Duties: Balancing Compliance with Agility Through Automated Controls and Tiered Risk

Segregation of duties (SoD) is a foundational control in traditional enterprise IT, deeply embedded in the organizational structure with its distinct Dev and Ops teams. However, the core principles of SRE—holistic ownership, build-run teams, and fluid collaboration—directly challenge the rigid application of SoD, necessitating a nuanced approach to balancing security, compliance, and operational efficiency.

In traditional legacy enterprises, the concept of **segregation of duties (SoD)** is fundamentally woven into the organizational fabric, creating a clear and often impermeable division: **"You are either Development or Operations."** This means you either **"write new features, build, configure, and integrate software,"** or you are exclusively **"operating software in production and have access to production data."** Your role dictates whether **"you can change the application or you have access to the application in production."**

This strict organizational construct, which dictates **"its own skillset, tools, and beliefs"** for each side, was primarily established for three main reasons:

- **Risk Reduction and Quality Control:** SoD embodies the **"checker and maker"** principle. One person or team performs a change, and another independent person or team controls and validates that change, intending to improve quality and reduce errors.

- **Fraud Prevention:** A core driver, particularly in financial services companies, is the prevention of **"fraud."** By requiring **"multiple people to change or operate an application,"** it becomes **"more challenging to manipulate and take advantages"** of systems, including sensitive production data, which, for some companies, represents their "key asset."

- **Regulatory Requirements:** Legal and compliance mandates often necessitate SoD. This can include data privacy requirements (e.g., restricting access to sensitive personal information like salary, handicaps, or worker council membership), which are legally required in certain countries. Implementing controls in the process and restricting access to systems are common means of compliance.

However, the very essence of an SRE role, and particularly the shift to **build-run teams**, inherently **"requires breaking this system"** of rigid SoD. An SRE aims to **"write code and also operate in production."** The ambition extends beyond individual SREs; it requires **"all team members to move fluently from writing code to taking over on-call and production support."** The core SRE tenet is that **"people who are changing the code also feel the change for the user when the code gets deployed to production."** This demands direct observation of the application and how their code performs in live environments, which directly challenges the corset that SoD traditionally imposes on practitioners.

The practical reality is that most SoD implementations in "highly regulated industries are not flexible enough for build-run teams." They often require cumbersome "time to move from Build to Run and the other way around," involving "approval," filling out "questionnaires," and receiving "limited time" access. This rigid, manual process is insufficient for the dynamic, "fluent" needs of SRE, where "people are required on both sites on the same day." The challenge, then, is to discover "how did other companies give less restrictions to their squads but also were able to keep up with their regulations? Or even improved audit control and reduced risk?"

This often requires engaging external or internal **auditors, technical specialists, and process consultants** to reimagine how SoD can be enforced through automated controls rather than manual, organizational separation. In legacy enterprises, specialized software enforces access control and tracks actions ("who has done what, when, and why"). Leveraging and enhancing these **"preventive and detective change controls"** is key.

The complexity extends across all IT layers: hardware (requiring physical access control), networks, middleware, operating systems, and applications. While the immediate focus for SRE often lies at the "application layer," a holistic review of all layers is necessary when changes are made to one. Clear communication about *which* layers are being modified is also vital.

Critically, decades of adherence to the traditional SoD model have ingrained a belief that **"quality only comes from supervision"**—from a **"Checker and Maker"** model. When transforming this model, it is essential to **"take care about clearing the belief that segregation of duties improves the quality."** While supervision *does* improve quality, **"we don't need a human to do that."** The SRE approach leverages **automation** to achieve the same or even higher levels of control and quality, without the inherent friction and delays of manual handoffs. This includes:

- Automated code reviews, static and dynamic analysis.

- Automated compliance checks within CI/CD pipelines.

- Automated deployment gates based on predefined criteria.

- Immutable audit trails generated by automated processes.

By shifting the focus of control from human-based organizational separation to **tool-based automated controls**, organizations can create a more agile and reliable environment that still satisfies stringent regulatory requirements. This requires a strong partnership between SRE teams, security, compliance, and internal audit functions to design and implement these next-generation controls.

For a truly effective **"build-it-run-it"** organization, the tangled "knot" of traditional segregation of duties (SoD) must be unraveled not by brute force, but by **meticulously defining "artifacts and quality gates" before unleashing automation**. This strategic approach acknowledges that simply throwing technology at a deeply ingrained organizational problem will fail unless the underlying process and control points are clearly understood and formalized.

Some enterprises attempted to address this with **"Zero-Access-to-Production"** during their DevOps transformations—the idea that **"no-one has access to production"** directly, with all changes occurring **"via authorized tools."** While seemingly a **"great idea,"** this concept often **"fails immediately when we put it into action"** in legacy environments. It might work for modern, greenfield cloud-native applications, but it requires a **"second concept for all the other applications with legacy technology that cannot adopt it."** The inherent complexity of diverse technologies within a legacy landscape means that **"each technology requires approved and audited deployment pipelines,"** which often don't exist or **"are not mature enough to provide all evidence for the audit log."** Furthermore, the need for direct access to production during critical outages, or for anonymizing production data for testing, often necessitates exceptions to strict zero-access policies. All of this, as we indicated earlier, **"requires time and investment."**

Our experience dictates a more pragmatic, **clustered approach**:

- **Cluster 1: Business Critical Applications:** For core, competitive advantage-driving applications, the aim is to allow experts "fluent and flexible moving between development and production." This requires **full automation of all activities**, which paradoxically "increases at the same time tracking" and enables robust auditability. The goal is to reduce fraud, enhance quality via scanning tools, and comply with regulations through hardened and audited tools. The caveat: this demands significant investment and highly skilled personnel. This approach also redefines what's "business critical," potentially elevating deployment tools to this status, as their failure impacts bug fixes.

- **Cluster 2: Low Business Critical Applications:** For applications whose failure only leads to minor, delayed complaints (e.g., a parking reservation app), a lighter touch is applied. For these, "mostly we need no segregation of duties," granting "full flexibility to SRE and build-run teams." The crucial learning here is to be "clear about what data is really stored," as even seemingly innocuous applications can contain sensitive, legally protected data (e.g., information on handicapped individuals).

- **Cluster 3: All the Other Applications:** This category often
  constitutes the majority of applications—those not critical enough
  for Cluster 1 investment but more impactful than Cluster 2. For these,
  elements of the Cluster 1 approach are applied where possible, while
  also initiating longer-term projects to address their challenges. This
  acknowledges that a perfect solution might never be found for every
  "Configuration Item in our CMDB" due to varied technologies and
  requirements. The challenge here is team motivation, as working on
  less-prioritized applications can be "demotivating." The solution is to
  use Cluster 1 as a "showcase" and provide support to Cluster 3 teams
  who wish to adopt its practices.

This clustered strategy accelerates transformation where possible while providing
targeted protection where required. It recognizes the heterogeneity of legacy IT and tailors
the SoD approach accordingly. To implement a robust segregation of duties (SoD) for **Cluster
1 (Business Critical Applications)** in a build-run model, the core concept is to enforce
control **through tools and automated pipelines**, rather than relying solely on manual,
human separation. This enables continuous delivery while ensuring auditability and risk
reduction. The principle is that **"each change to production requires at least 6 eyes,"**
representing three distinct roles, formalized and tracked within the automated workflow:

- **Change Requester (2 Eyes):** This is the individual or role (e.g.,
  Product Owner) who **"wants to change something in production."**
  The request itself, whether a new feature described in a change ticket
  (e.g., Jira) or a bug fix detailed in a problem or incident ticket (e.g.,
  ServiceNow), is meticulously **"documented"** and forms the initial
  "two eyes" of the process.

- **Implementer (2 Eyes):** This is the person who **"is implementing the
  change by updating the source code."** Tools like Git track every **"line
  of changed code,"** including author details and reasons, providing the
  next pair of "eyes" and an immutable record of the modification.

- **Reviewer and Activator (2 Eyes):** A crucial **"third person is
  required to actively and finally approve the change."** This role,
  typically a senior developer, architect, or SRE, provides the final "two
  eyes" by reviewing the code (e.g., through Bitbucket's documentation
  of feedback after inspection) and then *activating* the deployment.

The entire process is orchestrated through the **CI/CD pipeline**, which becomes the central mechanism for enforcing these controls and collecting audit evidence. All this data is **"stored immutable and not changeable,"** along with the formal Change ticket and all associated artifacts. This automated, **"6-eyes"** approach significantly **"reduces our audit time"** by providing comprehensive, verifiable evidence of **"who has done what and when."** It's a modern interpretation of SoD that balances high velocity with rigorous control, moving from human-dependent oversight to tool-enforced governance.

Accelerated Change Management: Automating Risk Assessment and Nudging Safe, Frequent Deployments

IT Change Management, in its essence, is the structured control mechanism for **"changes to assets, infrastructure, or software of an enterprise."** Its primary objectives—**securing quality, reducing fraud, and minimizing downtimes/breaks**—are strikingly similar to those of segregation of duties. While SoD is a concept, Change Management is a process. This process is absolutely **"essential for SRE,"** as **"minimizing downtimes and reducing outages starts with controlling what is changing and understanding what causes friction and outages."** SRE practitioners must not only be "well aware of management" to handle this volume. This creates an unsustainable bottleneck unless the process itself is fundamentally transformed through automation.

In traditional enterprises, the **IT Change Process is notoriously manual** and often a significant bottleneck. The workflow is cumbersome: a developer **"logs into ServiceNow, opens a ticket,"** then **"writes emails for approvals,"** and finally, the CAB **"approves the change."** The inherent flaw here is that **"people with no or low knowledge about the details of the change must give the approval and are held accountable for the change,"** leading to approvals based on limited understanding or **"personal knowledge"** rather than objective data.

This deeply manual process is fundamentally incompatible with the demands of modern, high-velocity IT. As has been observed, there's an increasing volume of changes, each smaller in scope and inherently less risky, and crucially, accompanied by a wealth of data about its quality and potential impact. This confluence of factors—more changes, smaller risks, more data—leads to a clear solution: **"We automate the manual**

work for low-risk changes." Higher risk forces teams to **"work more"** and discourages manual, untracked deployments. Three main factors influencing the risk of a change have been identified, which can be used to drive automation:

- **The Team Performing the Change:** A team with **"less experience is more likely to produce an outage."** Their historical **"track record of failed changes"** provides an objective measure of their performance.

- **The Application Itself:** Its **"business criticality"** (e.g., a trading platform for a bank) and its **"current situation"** (e.g., current instability, indicated by Error Budget or incident numbers) increase or decrease risk.

- **The Deployment Process:** A mature DevOps pipeline that produces **"tons of data in testing"** and leverages **"advanced release methods like canary deployment"** indicates lower risk compared to a **"manual"** copy of a build file into production.

By calculating a **risk score for each change** based on these parameters, an automated system can categorize changes as **"low-, medium-, or high-risk"** and **"act accordingly."** This allows for **"more automation and reduced controllers for low-risk changes,"** freeing up human reviewers to **"give more time and focus on high-risk changes."**

This data-driven approach also creates a powerful feedback loop for delivery teams. By understanding the **"risk classification and the influencing factors,"** teams are nudged to **"change their behavior and workflow to reduce the risk of their change."** This promotes continuous learning and the sharing of **"good behavior across teams to learn from each other."**

Customer-Centric Service Levels: Leveraging SLIs/ SLOs to Drive Real-Time, Preventative Action over Traditional Reporting

Service level management (SLM) is a deeply ingrained process in legacy enterprises, traditionally focused on defining, monitoring, and reviewing service indicators articulated in **service level agreements (SLAs)**. Its purpose is to assure customers that a required service is being delivered. However, SRE's approach to SLM represents a profound evolution, shifting from a reactive, after-the-fact reporting model to a proactive, preventative framework.

In traditional enterprises, SLM is primarily **"based on reporting the number of incident tickets, the solution time of incident tickets, and the number of reopened incident tickets."** While these metrics offer a snapshot of service performance, they are fundamentally **"after the fact"**: **"We count impacts when something already happened and assume that this measures our stability and realizability."** At best, this approach yields a **"one-dimensional"** view of service, a stark contrast to SRE's "shift... to prevention based on indicators" (traditional service levels often focus on whether an application is *up*; an SRE-driven SLO might focus on whether 99.9% of user login requests complete within 500 ms, or whether the shopping cart loads in under 2 seconds). These metrics are direct proxies for customer happiness and business outcomes.

While strict rules, like **"applications must show SLIs for latency,"** do not work due to the unique nature of each application and user journey, SRE still mandates that **"teams set up, monitor, and use the new set of SLI/SLOs."** Merely setting up SLI/ SLOs does not guarantee their effective use or the actions that follow. Key questions remain: "Who is responsible for acting when SLI/SLOs are not fulfilled?" and "Who must be informed and who takes actions when?" This proactive, centralized reporting allows leadership and domain owners **"to act before the fact."** They can **"see early enough how capacity planning and resource planning are impacting our customer."** Aggregated views across multiple applications (e.g., all sales applications or trading services) provide a holistic "pulse" of the business. SRE, for the first time, offers **"the option to act early enough and to act before the user experience is so impacted that we lose our reputation."** This transforms service level management from a compliance exercise into a strategic tool for proactive business resilience.

Quantifying the Unseen: Demonstrating SRE's Value in Traditional Companies

The implementation of SRE in a legacy enterprise, while challenging, yields a myriad of compelling benefits that directly impact both the technological landscape and the broader business outcomes. These advantages transcend the **"confidence to move**

strong in increasing the release frequency more and more." This confidence stems
from the inherent transparency SRE brings to how the system operates in agreement (via
SLO-driven SLAs) with the business. This cultivates a closer, joint understanding of how
technical debt might compromise system reliability, fostering a shared accountability
for balancing innovation with stability. This synergy ensures that IT is not merely a cost
center but a strategic enabler of business agility and customer satisfaction.

Beyond the strategic alignment and the shift in focus, SRE delivers concrete,
measurable improvements across operational efficiency and human capital.

- **Reduced Incidents:** SRE's proactive approach, leveraging real-time
  observability, chaos engineering, and automated risk assessments,
  significantly **reduces the number of incidents** in production. By
  identifying and mitigating weaknesses before they escalate, SRE
  moves organizations from reactive firefighting to preventative
  engineering.

- **Reduced Incident Severity:** When incidents do occur, SRE practices,
  such as intelligent alerting based on SLOs and rapid incident
  response protocols (e.g., automated runbooks), build-run teams
  foster **reduced tension and communication effort**, leading to a
  "10-15% reduction in effort."

- **Reduced Downtimes (MTTR):** SRE enables faster recovery from
  outages and minimizes the average time taken to resolve incidents
  and restore functionality.

- **Less Fluctuation (Improved Employee Satisfaction):** SRE cultivates
  a culture of **psychological safety**, where teams **"don't blame each
  other and thus feel more psychological safety,"** leading to faster
  learning and improvement in postmortems that drives tangible
  improvements in system reliability, operational efficiency, and
  ultimately, the well-being and productivity of the engineering
  workforce.

To demonstrate value and secure funding, SRE must be measurable and translated
into financial terms. Some teams start by counting incident tickets and calculating
perceived productivity savings in spreadsheets, with leadership often asking for specific
reductions in incident numbers. The high cost pressure in companies often drives this
focus. While teams have many success stories about saving time and reducing incidents,

these are often seen as "exceptional" or only applicable to unique situations like legacy tooling or missing observability. The sources recommend that instead of forcing top-down ticket reduction mandates, the focus should be on letting teams decide what makes the most sense for them, depending on their situation and needs.

However, for calculating Return on Investment (ROI) and building a business case, the number of incidents can be translated into financial impact. Two main options are available for this calculation (**calculate based on the number of incidents and calculate based on addressed issues) as detailed later in Chapter 7.** These metrics are not just numbers; they help develop a strategy to track improvements brought by SRE at each stage of the capability implementation. With a solid understanding of these KPIs, organizations can further enhance efficiency and streamline processes by reducing SRE's toil.

The ultimate goal of SRE is to construct, operate, and continuously improve systems built for real-world demands, ensuring future-ready reliability where systems evolve, self-heal, and drive innovation at scale.

Summary

This chapter comprehensively illustrates how **SRE principles** offer a blueprint for profound organizational and cultural transformation within traditional IT landscapes. The chapter critically examines the **enduring divide** between development ("build") and operations ("run"), highlighting a pervasive structural schism that fosters **fragmented responsibility, externalized quality control, and deep-seated mistrust**. This traditional model inherently lacks **proactive, measurable insights into IT stability**, relying instead on "after-the-fact" incident metrics, which shield developers from the direct "pain" their changes are causing the customer. Furthermore, a **fear-driven mindset** and rigid entitlements have left operations teams "untouched over years" by modern agile and DevOps transformations, creating a "two-speed IT organization." To bridge this chasm, SRE advocates for **building "build-run teams,"** fundamentally reshaping team structures and organizational dynamics. The transformation, however, is not a complete rewrite but a **strategic "update, change, and adjustment" of the existing operating model**, guided by SRE epics such as "Measuring Everything," "Leverage Tooling for Automation," "Accept Failure as Norm," and "Simulate and Test." The **"You Build It, You Run It" philosophy** becomes the cornerstone, necessitating alignment of leadership objectives equally valuing stability and speed and leveraging

Error Budgets, SLIs, and SLOs to make customer experience transparent and reliable as a data-driven business decision. This involves strategically integrating existing "operations experts into the squad" through the "people follow function" imperative, while addressing impediments like worker contracts, developer resistance to support, and skill gaps through targeted upskilling and a focus on reducing operational burden. SRE also dramatically **modernizes operational processes**, shifting incident and problem management from a reactive "after-the-fact" approach to a proactive, anticipatory one, integrating incident response into application teams' daily rhythm via a **team rotation model** and the "Observe, React, Solve, Learn Method." Problem tickets are no longer left to languish but become prioritized backlog items, with **Error Budget** guiding the balance between new features and stability improvements. A critical emphasis is placed on **"continuous automation"** to reduce "toil," and **segregation of duties (SoD)** is reimagined by enforcing controls through **automated tools and pipelines** rather than manual separation, accelerating IT Change Management based on calculated risk scores. Service level management evolves into a customer-centric framework, enabling teams to act *before* user experience is impacted. Ultimately, SRE's value is **concrete and measurable**, providing the confidence for **increased release frequency** and demonstrating tangible benefits like reduced incidents, decreased severity, shorter mean time to restore (MTTR), and improved employee satisfaction through psychological safety. This value can be translated into financial terms using key performance indicators (KPIs) such as cost per incident, system availability, and change fail rate, transforming reliability from a cost center to a **strategic competitive advantage**.

CHAPTER 4

Culture Shift for SRE Adoption

In this chapter, we explore the significance of culture within site reliability engineering (SRE) and its pivotal role in fostering resilient and scalable systems. Culture is not merely a backdrop; it acts as the foundation that guides successful practices in SRE. We will discuss the principles of collaboration and sharing, emphasizing how a culture that promotes knowledge sharing among teams is vital. This openness dismantles silos that can separate development, operations, and SRE functions, facilitating faster problem detection and resolution, which ultimately leads to more reliable services. Additionally, we will examine the concepts of empowerment and psychological safety, highlighting the importance of trusting team members to make decisions and take ownership of their work. This trust, coupled with accountability, allows SRE teams to take informed actions and drive meaningful change. Finally, we will address the necessity of continuous learning in SRE. As the field evolves, so must the teams; fostering a culture that encourages curiosity, experimentation, and sharing of insights is crucial for keeping pace with ever-complex challenges. Through these discussions, we will see how nurturing these cultural facets is essential for cultivating an effective SRE mindset.

This chapter will cover the following main topics:

- **Why Culture Matters in SRE**

- **Principles of Collaboration and Sharing:** A culture of sharing is indispensable for successful SRE. When knowledge, experiences, and best practices flow freely across teams, it fosters mutual understanding and unity. This open sharing breaks down silos that commonly form between development, operations, and SRE groups. It also enables quicker detection and resolution of issues, leading to more resilient services.

- **Empowerment and Psychological Safety:** These involve trusting people to make decisions, solve problems, and take action. They go hand in hand with accountability. SRE teams must feel they have both the authority to implement changes and the responsibility for the outcomes.

- **Encouraging Continuous Learning:** SRE is not static; it demands that teams constantly refine their skills and adapt to evolving challenges. A culture of continuous learning encourages engineers to stay curious, experiment, and share their findings.

Why Culture Matters in SRE

Culture plays a crucial role in SRE because it shapes how teams think, collaborate, and innovate. SRE is not merely a technical discipline; it's fundamentally about a mindset that emphasizes reliability, resilience, and a proactive approach to problem-solving. A significant aspect of an SRE's work involves empowering others within the organization to adopt this SRE mindset. This consists in nurturing a culture of shared responsibility and encouraging teams to think critically about engineering solutions that are scalable and sustainable. Ultimately, SREs help lay the groundwork for building future complex systems by guiding teams to effectively utilize engineering principles to enhance overall operational performance and reliability. The importance of cultivating the right culture cannot be overstated; neglecting it can lead to significant setbacks. For instance, Netflix's notable 2016 outage highlighted how a lack of clear communication and siloed knowledge between teams delayed incident resolution, while Facebook's global outage in October 2021 illustrated the consequences of insufficient resilience planning and chaos engineering practices. Both incidents underscore how crucial a collaborative, transparent, and proactive reliability culture is to maintaining robust, resilient systems.

Principles of Collaboration and Sharing

In this section, we delve into **transparency**, **collaboration**, and **community ownership** to show how an open exchange of knowledge and responsibilities underpins a supportive environment for achieving reliability goals.

- **Transparency:** Distribute project objectives, incident learnings, and performance data so every team can work from the same information and align on shared goals.

   Example KPI: Percentage of incident postmortems documented and shared within 48 hours of resolution.

- **Collaboration:** Embed SRE principles into day-to-day workflows, encouraging different teams—from development to operations—to partner closely.

   Example KPI: Average time taken to detect and resolve incidents (MTTD/MTTR), measured before and after implementing cross-team collaborative practices.

- **Community Ownership:** Cultivate a sense of collective responsibility for reliability, ensuring that no single team or individual shoulders the burden alone.

   Example KPI: Ratio of incidents resolved by cross-functional teams versus those handled by single-team silos

In SRE, culture is the foundation upon which successful teams operate. A team culture must be well embedded in the firm's culture. It shapes how teams collaborate, solve problems, and automate smartly by sharing knowledge and continuously learning from each other, making it as critical as the technical systems they manage.

The SRE culture we want to activate revolves around principles that promote innovation, resilience, and trust within the team and between the leadership and the teams. This will drive further empowerment so that teams can work independently. From embracing failure as a learning opportunity to fostering data-driven decisions, these values empower teams to manage complex systems with confidence.

By prioritizing a culture of collaboration and accountability, organizations that practice SRE have transformed their approach to reliability, achieving results such as faster recovery times, fewer incidents, and better team alignment.

This chapter explores the essential components of SRE culture and the behavior we want to foster. It explains how principles like shared responsibility, blameless retrospectives, and a focus on measurable outcomes create an environment where teams thrive and systems excel.

We structure behavior and culture in five main areas. During our transformation, we always focus on one area and combine it with our SRE capabilities.

Knowledge-Driven Collaboration

SRE culture fosters **open communication** and **collaboration** between teams. Open communication and shared resources like runbooks and playbooks (Example: Listing 4-1) **reduce silos** and **enable consistency, for example, in incident response.** Standardization brings us to a position where we can learn from each other faster. But at the same time, we want to challenge the current status and innovate how we work. This needs to be balanced and discussed. We want to accelerate sharing by offering forums to share.

Listing 4-1. Playbook example with inline documentation call-out to enhance transparency

```
# example ansible playbook fix checkout service kubernetes deployment
---
- name: Ensure checkoutservice deployment has always 3 replicas
  hosts: kubernetes_prod
  gather_facts: no

  tasks:
    # Step 1: Get current replica count for checkoutservice
    - name: Check current replicas for checkoutservice
      kubernetes.core.k8s_info:
        kind: Deployment
        namespace: default
        name: checkoutservice
      register: checkout_deploy

    # Step 2: Patch deployment to 3 replicas if needed
    - name: Fix replicas to 3 if not already correct
      kubernetes.core.k8s:
        kind: Deployment
        namespace: default
        name: checkoutservice
```

```
      replicas: 3
      state: present
    when: checkout_deploy.resources[0].spec.replicas != 3

  # Step 3: Verify replicas are now 3
  - name: Verify checkoutservice replicas after patch
    kubernetes.core.k8s_info:
      kind: Deployment
      namespace: default
      name: checkoutservice
    register: verify_checkout

  - name: Show final replica count
    debug:
      msg: "Replicas for checkoutservice: {{verify_checkout.resources[0].
spec.replicas }}"
```

For example, in the community of practices, during training, and in joint places on the intranet, cooperation can take place.

Collaborative knowledge-sharing fosters a culture where teams learn from failures and explore risks safely. Blameless post- and pre-mortems are enriched by openly discussing insights and risks, creating a psychologically safe space.

For example, we could conduct pre-mortem exercises, during which teams collaborate to identify potential failure points and share mitigations. This would improve incident prevention.

The capabilities in focus are

- Blameless postmortem

- Pre-mortem/what-if scenario

- Psychological safety

And we want to enable with

- Meeting cadences to share success stories

- Trainings and certificates to learn together

- Standardization in technology and processes

- Collaboration tools and virtual places

Smart Automation

To develop and architect systems so that they can scale independently from people doing operations, we want our practitioners to think about opportunities they can automate constantly. But we smartly wish to do that. Not each manual task is ready for automation. We want our tools, scripts, and platforms to be built to **manage routine operational functions** so that we enable our teams to focus on higher-value engineering tasks. That's important because automation **reduces** human error, **accelerates** response times, and **frees** up engineers to tackle complex, high-impact tasks.

The culture of thinking "how can I smartly automate that" is combined with our core SRE capabilities:

- Toil management

- AI Ops

- Automated on-call support

For example, we use AI Ops (Splunk ITSI, BigPanda, Dynatrace, etc.) to intelligently triage incidents during on-call support, automating common resolutions and minimizing toil. We can also use analytics to improve and automate the risk assessment for IT change tickets.

Collective Reliability Mindset

SREs bridge the gap between development and operations, ensuring reliability is a core part of the software delivery lifecycle (SDLC). In the SDLC, more teams must collaborate and finger-pointing is common; focusing only on optimizing one's own silo no longer works. What the teams need is to share the responsibility for the customer experience. That the system delivered as promised. Reliability cannot be siloed; it requires alignment between business, development, testing, infrastructure, platform, and operations. This starts with clearly defining reliability in user terms.

Because each team may meet its own SLOs and still appear "green" internally, the **only reliability that matters is the user's experience**, and this must be visible to everyone within the organization. For example, we can define a simple **user-centric service level indicator (SLI)** as

- In Banking customer—Example Credit Card Transaction

    Successful Transaction Rate: *SLI = (Successful Credit Card Transactions / Total Credit Card Transactions) × 100%*

- In Retail customer—Example Checkout Service

    Cart to Order Conversion Reliability:

    SLI = (Orders Placed Successfully / Checkout Attempts) × 100%

- In Automotive—Example Dealer Service Booking

    Successful Booking Service Rate

    SLI = (Service Bookings Placed Successfully / Total Booking Attempts) × 100%

A shared commitment to reliability ensures that SLOs and error budgets are treated as team goals, driving alignment across stakeholders. Collective ownership benefits capacity management and resilience design patterns, ensuring that reliability considerations are embedded at every layer.

Our core capabilities, which we enroll with our collective reliability mindset, are

- Capacity management

- SLOs and error budget

- Resilience design patterns

Empowerment and Psychological Safety

We detail decentralized decision-making (Overview: Figure 4-1), clear accountability, and support for innovation to illustrate how giving teams autonomy and responsibility fuels faster decision-making and drives meaningful ownership of outcomes.

- **Decentralized Decision-Making:** Enable those on the front lines to respond swiftly to incidents or performance issues, reducing bottlenecks and fostering agility.

- **Clear Accountability:** Define ownership of services or features so everyone knows who is responsible for what and can act confidently.

137

- **Support for Innovation:** Provide the psychological safety and resources for individuals to experiment, iterate, and learn from both successes and setbacks.

Figure 4-1. *From Bottlenecks to Full Autonomy, Empowering Teams Step by Step*

Data-Informed Decisions

Change must always be driven by evidence. We want to have metrics to show improvements and to make decisions in the right way. Empowerment requires trust, and we can articulate status, improvements, and decisions. It is best if we measure everything and be scientific about that—metrics-driven decisions. We use metrics over assumptions—for that, we track, analyze our hypotheses, and make decisions. It is best to centrally collect SDLC and ITSM data to optimize IT processes by comparing efforts and approaches to each other. Monitor the system with health data at the team level for our operational decisions. A central metric store helps in standardization and allocation of funding for more significant optimizations on an enterprise level, as well as helps to identify bottlenecks, track progress, and validate optimizations. For example, teams can use data-driven approaches to make informed decisions about system changes, leveraging synthetic monitoring and feature flags (Example: Listing 4-1) to validate changes safely. Risk assessment tools use data to evaluate potential impacts proactively. We can use feature flags to release updates gradually, monitor user impact via synthetic monitoring, and adjust based on data-driven risk assessments.

Listing 4-2. Feature flag example definition JSON

```
{
  "feature_name": "winter_sunday_sale_banner",
  "enabled": true,
```

```
  "description": "Show a promotional banner on homepage for Winter
  Sunday Sale",
  "rollout": "100%",
  "created_at": "2025-06-30T00:00:00Z"
}
```

Listing 4-3. Feature flag example to activate promo link by query to
PostgreSQL table

```
...
# Example function to get JSON flag
def get_feature_flag(flag_name):
    conn = psycopg2.connect(...)
    cur = conn.cursor()
    cur.execute("""
        SELECT json_build_object(
            'feature_name', feature_name,
            'enabled', enabled,
            'description', description,
            'rollout', rollout,
            'created_at', created_at
        )
        FROM feature_flags
        WHERE feature_name = %s
        LIMIT 1
    """, (flag_name,))
    result = cur.fetchone()
    cur.close()
    conn.close()
    return result[0] if result else None

# Evaluate in production code
flag = get_feature_flag('winter_sunday_sale_banner')
if flag and flag['enabled']:
    ... (code when enabled)
else:
    ... (code when not enabled)
```

This core capability goes beyond SRE but is so critical for our success:

- Developer experience

- Central metrics store/value office

- Synthetic monitoring

- Feature flags

- Automated risk assessment in change management

Embedding Culture As the Catalyst for SRE Success

Culture is the unseen force that defines how organizations approach challenges, innovate solutions, and build resilience. In SRE, the behavior of intelligent automation, knowledge-driven collaboration, a collective reliability mindset, blameless feedback, and data-informed decisions are not just theoretical—they are the building blocks of sustainable reliability. They will reduce the adoption time during our transformation.

Activating this culture requires more than slogans or directives; it demands a concerted effort to align behaviors, values, and technical capabilities. Leaders must model the culture they want to instill, empowering teams to embrace responsibility and failure as opportunities for growth. For example, leaders can share when they failed or when they were at a point in their careers where not everything was running as planned. Leading by example will activate and accelerate their journey.

As we continue our SRE transformation, we will focus on embedding these cultural principles into every layer of the organization. Ask yourself:

- How can we foster a psychologically safe environment where experimentation thrives and failures are seen as learning opportunities?

- Are we empowering teams to think beyond their silos, working collectively toward reliability as a shared outcome?

- What steps can we take to ensure our decisions are consistently informed by data and aligned with measurable outcomes?

Encouraging Continuous Learning

SRE evolves rapidly, and maintaining excellence requires an ongoing commitment to learning. An organization that values continual improvement encourages engineers to stay inquisitive, adopt fresh perspectives, and actively share their discoveries.

In this section, we discuss **Blameless Postmortems**, **Training and Development**, and **Responsive Feedback Loops** because each mechanism reinforces a growth mindset, ensuring that lessons learned from successes and failures alike are consistently applied to improve systems and processes.

- **Blameless Postmortems**: Use incidents as educational moments, focusing on identifying root causes and preventing recurrences rather than pointing fingers.

- **Training and Development**: To keep pace with emerging technologies, offer regular opportunities for skill-building, such as workshops, certifications, or mentorship.

- Example learning path for DevOps engineers, new SREs, and software engineers:

  Week 1: Site Reliability Engineering Foundation (DevOps Institute or Udacity or any other equivalent)

  Week 2: Art-of-SLO (Google SRE), Observability Foundation (DevOps Institute or equivalent), and OpenTelemetry

  Week 3: Psychological safety, blameless culture, and chaos engineering (available on Coursera or Edx or Udacity)

  Week 4: Software engineering principles and DevOps practices (CI/CD and release management)

  Weeks 5–6: Kubernetes, Ansible, network fundamentals, and IaC

- **Responsive Feedback Loops**: To refine practices and processes, collect and integrate feedback from various sources, such as monitoring tools, user suggestions, and retrospective reviews.

Blameless Postmortems Feedback Loop

To be responsible for reliability starts with accepting that we all are producing failures in our work. Teams should not get laid off because of failures. We want to encourage our people to speak about failures that we all can learn from. We need to create a blameless culture. A culture of learning from failures without assigning blame to individuals, but also planning for failure. A blameless culture fosters a culture of trust and learning. Teams can address the root causes of incidents constructively. Blameless feedback ensures that chaos testing and reliable experiments lead to actionable insights without fear of blame. Observability tools help close the loop by surfacing data that informs continuous improvement. For example, after a chaos engineering experiment reveals weaknesses, teams analyze results collaboratively, updating observability dashboards to prevent similar issues.

The capabilities in focus are

- Psychological safety

- Blameless postmortems

- Chaos experimentation and engineering

- Reliable experiments

- Observability

Summary

The cultural transformation outlined in this chapter is not a one-time initiative; it's ongoing. By prioritizing these behaviors and combining them in our transformation with capabilities, our organizations unlock the full potential of SRE. In this way, we will not demand too much from our teams. We are changing step by step and focusing on a cultural aspect with the relevant capabilities. We create systems that not only perform reliably but also inspire confidence and trust across teams and stakeholders alike.

Transforming an organization's culture along these lines is crucial to thriving in SRE. By embracing open communication, granting engineers the autonomy they need, and championing continuous improvement, teams can deliver high-quality, resilient services in fast-changing environments. When everyone is aligned on these cultural pillars, the result is a cohesive, motivated, and adaptive organization—one poised to deliver on the promise of true reliability.

The next chapter will dive deeper into **the SRE transformation**, building upon this cultural foundation to explore the tactical and strategic aspects of implementing SRE in your organization.

## CHAPTER 5

Essential Skills for SRE Practitioners

In this chapter, we embark on a deep exploration of the competencies that transform a competent engineer into an SRE powerhouse, beginning with the craft of technical proficiency—where writing elegant code and sculpting automation pipelines becomes second nature, and Infrastructure as Code and CI/CD practices forge environments that heal themselves under pressure. We will discover how the precision of our scripts and the resilience built into every deployment not only slays toil but also lays the foundation for rapid, reliable change. As we shift focus, we will learn to wield communication as a strategic tool—narrating vivid stories through SLO-driven dashboards, translating complex system behaviors into executive-ready insights, and forging alliances across development, security, and leadership teams so that every reliability initiative feels like a shared victory.

When the unexpected strikes, we will guide you through high-stakes incident response techniques that sharpen the instinct for rapid triage, reduce MTTD and MTTR, and convert every outage into a springboard for systemic strength via blameless postmortems. Finally, we'll cultivate the change agent mindset that drives continuous improvement at scale, empowering you to spread reliability best practices across the organization, champion a culture that celebrates automated guardrails, and turn resistance into collective ownership of uptime and performance. By the end of this journey, we won't just master SRE skills; we will lead a transformational shift in how the business views reliability, embedding operational excellence into its very DNA.

This chapter will cover the following main topics:

- Technical proficiency areas (coding, automation, DevOps, observability, etc.)

- Communication and influence skills

F. Hoeppner and F. Sbaraglia, *Mastering Site Reliability Engineering in Enterprise*,
https://doi.org/10.1007/979-8-8688-1448-8_5

- Problem-solving and incident response

- Change agent mindset

Introduction to SRE Mindset

Before diving into the nuts and bolts of engineering, code, and automation, it's crucial to recognize that accurate site reliability engineering extends far beyond technical prowess.

> *SRE, or* Site Reliability Engineering, *is when you task a software engineer to design and manage an operations function. It's a software engineering approach to operations, focused on building and maintaining reliable, scalable, and highly available systems. Instead of traditional operations teams, SRE teams use code and automation to manage and improve production systems.*
>
> —Definition from the Google SRE book [https://sre.google/sre-book/introduction/]

At its core, SRE is as much about people and culture as it is about systems and services. SREs are the ambassadors who bridge development, operations, security, and business teams—cultivating shared ownership of reliability and fostering a culture where collaboration, transparency, and empathy drive continuous improvement. By mastering the art of influence, communication, and change management, SREs empower cross-functional teams to embrace automated guardrails and resilience-first mindsets. In this way, the SRE not only engineers technical solutions but also engineers organizational change, paving the way for sustainable innovation and collective success.

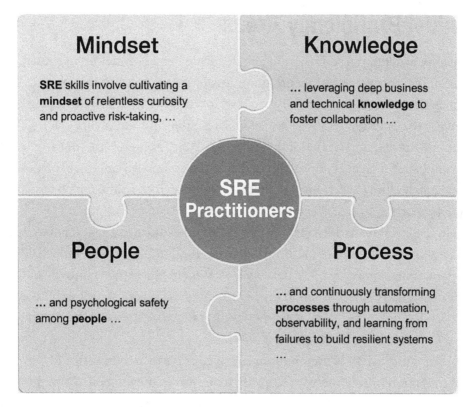

Figure 5-1. *Overlapping Diagram Across Mindset, People, Process, and Knowledge*

Site reliability engineering isn't just about lines of code and software engineering; it also demands soft **skills**, deep business **knowledge**, and the right **mindset** as presented in Figure 5-1. For example, an SRE's knowledge of how SRE principles map to business goals—embodying customer values and tailoring reliability practices to industry context—is just as crucial as any technical skills. Their skillset shines when designing advanced system architecture, observability pipelines, and automation workflows—using techniques like distributed tracing, log correlation, and self-healing scripts to reduce toil, resolve incidents in record time, and leave tomorrow's systems more robust than today's. And above all, the SRE mindset, fueled by curiosity, drives them to probe every failure, unite diverse perspectives as practice ambassadors, and foster a growth-oriented culture of collaboration across DevOps, Ops, ITIL, and Agile teams.

Technical Proficiency Areas

At the heart of exceptional site reliability engineering lies a foundation in software craftsmanship: fluency in multiple programming languages and an instinct for clean, maintainable code. SREs don't merely wield scripts as fire extinguishers during emergencies; they partner with product teams from day one, authoring features, contributing libraries, and embedding resilience patterns directly into the application code. By understanding every branch and dependency, an SRE can trace the flow of data, anticipate failure modes, and dramatically shorten mean time to detection (MTTD) and mean time to repair (MTTR).

Equally critical is deep domain knowledge. Whether the organization operates a global payments platform or a real-time video service, familiarity with industry-specific pain points empowers it to bake simple, targeted safeguards into the earliest design discussions. An alert tuned to the heartbeat of the business, rather than generic thresholds, becomes a precision instrument rather than background noise, surfacing only the anomalies that truly matter.

Beyond code, mastery of automation, IaC (e.g., Terraform, Pulumi), and DevOps practices transform how reliably we ship. SREs design self-healing pipelines that automatically provision, configure, and patch environments, eradicating manual toil and ensuring consistency across every deployment. Continuous Integration and Continuous Deployment (CI/CD) (e.g., GitLab, GitHub Actions, etc.) frameworks become the arteries of infrastructure, delivering changes rapidly yet safely, with built-in validation gates that catch regressions before they ever reach production.

Finally, observability and monitoring form the compass that guides every reliability journey. Instrumenting applications to emit rich, structured telemetry—traces, metrics, and logs—gives the granular visibility needed to diagnose issues under real-world load. More than simple "up/down" checks, true observability surfaces the internal state of distributed systems, helping to pinpoint the precise microservice, database query, or network hop at fault.

Together, these technical proficiencies—software engineering, domain expertise, automation, DevOps, CI/CD, and observability—create a virtuous cycle of insight and action. By embedding them into every phase of the lifecycle, SREs not only safeguard uptime but also accelerate innovation, turning reliability into a strategic engine for growth.

Communication and Influence Skills

At its core, site reliability engineering thrives on clear, empathetic communication that bridges the gap between technical trenches and executive boardrooms. An SRE must translate raw telemetry into compelling narratives, turning dashboards and error-budget reports into stories that resonate with developers, product owners, and C-suite leaders alike. By presenting service-level indicators alongside business impact metrics, this will help to foster a shared understanding of how reliability underpins revenue, customer satisfaction, and brand trust.

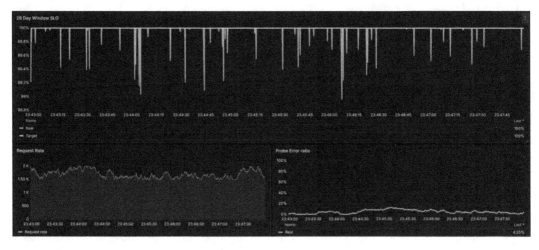

Figure 5-2. *Example of SLO Dashboard Checkout Service*

Influence in SRE emerges from building genuine partnerships across teams. Whether the team is running blameless postmortem reviews or facilitating cross-functional workshops, the ability to listen actively, ask the right questions, and surface diverse viewpoints creates a culture of mutual respect. When the team invites security, development, and operations stakeholders into reliability discussions to empower them to co-own SLIs and collaborate on solutions, it shifts reliability from a siloed responsibility into a collective triumph.

One-Page SRE Business Case

When making the case for strengthening site reliability engineering (SRE) in an organization, it helps to keep the proposal simple, clear, and focused on what leaders care about most: **customer trust**, **cost of downtime**, and **ROI**.

A good one-page business case starts with context:

- *What's the problem?* For example, frequent incidents, slow recovery, or an unreliable platform can erode customer confidence and increase costs.

- *What's the opportunity?* Investing in SRE practices can systematically reduce downtime, improve incident response, and help the business deliver new features faster and more safely.

Next, illustrate the **customer impact**. Quantify current pain points like average downtime hours per year and the cost per hour of an outage. Then set a clear goal, such as cutting downtime by 50% to protect revenue and keep customers loyal.

A simple ROI snapshot makes the benefits tangible. For example:

- Reduce downtime from 20 hours to 10 hours per year, saving hundreds of thousands of dollars.

- Cut change failure rates in half, which means fewer rollbacks and hotfixes.

- Improve customer trust, measured by higher NPS or retention.

The **investment ask** should be easy to digest: for example, adding two SRE engineers, introducing a chaos engineering tool, or upgrading monitoring capabilities— with an estimate of annual costs and a payback timeline.

Finally, close with next steps: get executive sponsorship, kick off hiring or training, and track progress with clear metrics like SLO attainment, change failure rate (CFR), and mean time to recovery (MTTR).

By summarizing the need, impact, and return on a single page, you make it easy for stakeholders to say **yes** to reliability—because trust is always worth it.

Negotiation and persuasion become essential tools when facing competing priorities. Crafting concise, data-driven proposals, such as justifying investment in automated rollback scripts or enhanced tracing, helps to secure the resources and buy-in needed to harden the systems. By weaving customer anecdotes, incident learnings, and ROI forecasts into the presentations, abstract technical debt can be turned into tangible business cases that executives champion.

Ultimately, an SRE's communication prowess lies in humility and curiosity. Embracing feedback, adapting the style for different audiences, and continuously refining the messaging strengthen the role as an ambassador of reliability. Through thoughtful dialogue and strategic storytelling, an SRE can transform emergency fire drills into collaborative problem-solving sessions, ensuring that every voice is heard and every improvement becomes a shared victory.

Problem-Solving and Incident Response

Problem-solving in SRE begins with a spirit of relentless curiosity—an instinct to probe beyond surface symptoms and trace every anomaly back to its root cause. When an unexpected latency spike or error surge emerges (Netflix's Chaos Monkey), an SRE doesn't stop at "the database is slow"; they methodically gather traces, correlate logs, and replay traffic patterns to recreate the precise conditions that triggered the issue. This investigative rigor uncovers not only what failed, but why it failed—and illuminates latent weaknesses before they bloom into outages.

To sharpen that investigative instinct, SRE teams employ the "What If?" methodology—regularly asking, "What if this service loses its primary database?" or "What if a sudden traffic burst hits this endpoint?" By simulating these scenarios in safe environments, new and seasoned engineers alike cultivate the curiosity and hands-on skills needed to dissect incidents under pressure and anticipate future failures.

Sustainable incident response is about more than urgent firefighting; it's a framework built to endure. By crafting blameless postmortems (e.g., Github repository templates `https://github.com/dastergon/postmortem-templates`) that live in a shared knowledge base, SREs transform each incident into a teachable moment—complete with timelines, decision logs, and actionable remediation steps accessible to every team across the enterprise. Furthermore, implementing a "follow-the-sun" response model ensures 24/7 coverage without burning out a single engineer. When one region's shift ends, another team seamlessly picks up the lead, keeping nighttime interruptions minimal and preserving long-term team well-being. This is also the methodology that helps when SRE needs to investigate complex issues that require a global team to run the investigation.

Proactivity underpins both preparation and continuous improvement. Regular chaos engineering exercises—whether deliberately throttling network links or injecting service failures—surface hidden coupling and configuration drift, turning unknown unknowns into documented improvements. For new joiners, these drills become the daily bread that hones diagnostic instincts; for senior SREs, they spotlight high-impact optimizations that drive MTTD and MTTR ever lower. In this way, incident readiness evolves from a reactive scramble into a confident, collaborative rhythm that sustains reliability and team happiness over the long haul.

Change Agent Mindset

Embracing the change agent mindset means viewing every reliability challenge as an opportunity to reshape both technology and culture. The SREs are the catalysts for continuous improvement, identifying friction points in workflows and embedding automated guardrails that nudge teams toward best practices. By collaborating early with product managers, architects, and security experts, they can influence design decisions that eliminate toil before it starts and bake resilience into every feature.

Change agents speak the language of empathy and data. They learn to listen to developer pain points, whether it's a slow build pipeline or opaque error logs—and translate them into prioritized projects that deliver visible impact. Armed with metrics from chaos experiments and postmortem analyses, they craft persuasive cases for investing in observability enhancements or refining deployment pipelines. This blend of technical insight and strategic storytelling turns skeptics into sponsors, forging alliances that carry reliability initiatives forward.

Ultimately, the SRE as a change agent is the steward of a growth-oriented culture. This will champion knowledge sharing through brown-bag sessions, documentation sprints, and reliability-focused hackathons—creating forums where every engineer can learn new techniques and contribute innovations. By celebrating small wins, recognizing cross-team collaborations, and surfacing lessons from failures without blame, they foster an environment where experimentation thrives and continuous learning becomes the norm. In this way, the SRE not only fortifies systems but also transforms organizations into adaptive, resilient enterprises.

SRE Adoption Roadmap: From Awareness to Practice

Figure 5-3 presents a visual roadmap outlining the progression of SRE adoption—from initial awareness to mature, resilient engineering practices. This journey is not a linear checklist but a structured evolution that organizations undergo as they embed SRE culture, principles, and operations into their technical and organizational DNA.

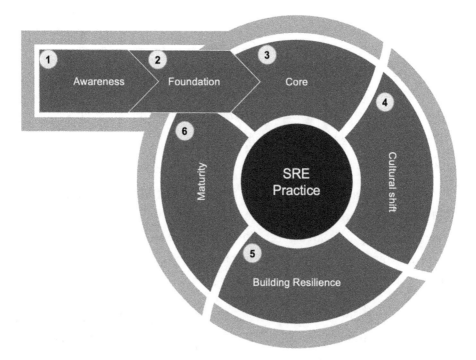

Figure 5-3. *Building SRE Capability from Awareness to Practice*

Phase 1: **Awareness**

The starting point is creating awareness by introducing core SRE principles and aligning them with existing DevOps practices. This foundational step builds the initial understanding of reliability as a discipline and sets the stage for deeper integration.

Phase 2: **Foundation**

In the foundational phase, teams begin implementing the basic building blocks of SRE, such as observability, alerting, and incident response. Ownership over services starts to shift to the engineers managing them, cultivating accountability and reliability thinking.

Phase 3: **Core SRE**

At this stage, organizations define critical user journeys, service level indicators (SLIs), and objectives (SLOs), begin automating operational tasks, and establish a sustainable on-call model. These efforts solidify reliability as an engineering responsibility and reinforce data-driven decision-making.

Phase 4: **Cultural Shift**

SRE becomes a catalyst for cultural transformation. Teams adopt blameless postmortems and embed reliability into planning cycles. This shift creates psychological safety and prioritizes learning over blame, essential for innovation and resilience.

Phase 5: **Building Resilience**

Advanced practices like proactive failure injection and chaos experimentation are introduced. These exercises are used not just to validate system behavior under stress but also to build team confidence and uncover hidden risks before they manifest.

Phase 6: **Maturity**

In the maturity phase, organizations adopt an engineering-first mindset—automating everything that can be automated, enforcing error budget policies, and conducting regular cross-team reliability reviews. This stage emphasizes continuous improvement, collaboration, and sustainable operations.

These phases often **overlap and repeat** as organizations grow; teams may revisit earlier stages to deepen practices, refine cultural alignment, or adapt to new challenges. SRE is an iterative journey rather than a rigid, one-way path.

The circular core labeled "SRE Practice" illustrates that SRE is not a destination but an ongoing cycle of learning, adaptation, and growth. Each phase builds upon the one before it, like stones forming a strong pyramid. Without a solid base, later stages become fragile. That's why each phase must be reliable in itself, with strong foundations that support the next layer of maturity. The strength of an SRE team's resilience depends on the depth and reliability of its foundations—technical, procedural, and cultural. By investing in each phase with care, organizations build not only reliable systems but also a sustainable and adaptable SRE culture.

SRE Production Readiness

In closing, the accurate measure of an SRE team's strength lies not only in how it responds to the crisis but in how consistently it prepares for the next one. By running full-suite readiness reviews, combining incident simulations, chaos experiments, skills workshops, and "What If?" tabletop exercises—every three to six months, to keep diagnostic instincts sharp, validate process improvements, and reinforce cross-team collaboration. This regular cadence transforms readiness from a one-off event into an ingrained discipline, ensuring that when real outages occur, the team steps forward with confidence, cohesion, and the practiced agility to keep services running flawlessly.

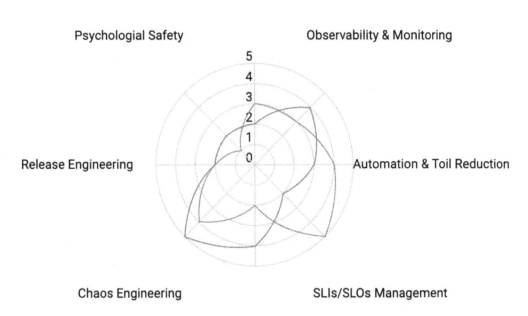

Figure 5-4. *Example of SRE Production Readiness Team A and Team B*

Summary
From Skills to Impact

In this chapter, we explored the essential skill set that defines an SRE practitioner, not just as a technical contributor, but as a force for resilience and cultural transformation. We began by honing the technical capabilities that reduce toil and increase system robustness, from clean code and infrastructure as code to CI/CD pipelines that enable safe, automated change. We highlighted how observability and intelligent automation are not just tools but cornerstones of self-healing systems.

Beyond code, we uncovered the strategic power of communication, how SREs bridge the gap between deep system insight and executive understanding, using SLOs and storytelling to build cross-functional trust and shared ownership of reliability.

Improving Stability	Solving Incidents	Prevention
<ul><li>Toil Management</li><li>Automated Infrastructure provisioning</li><li>Resilience Design</li><li>Continuous CI/CD</li><li>Cloud Costs Management</li></ul>	<ul><li>On-Call Management</li><li>Patching, Eng-point security</li><li>Cloud Backup/Disaster Recovery</li><li>Build-run Rotation</li><li>Runbook Automation</li><li>Drift Detection and Reporting</li></ul>	<ul><li>SLI/SLOs for Golden Signals with Error Budget</li><li>Monitoring and Observability</li><li>Release Strategies and Feature Flags</li><li>Chaos Engineering</li><li>Blameless Post Mortems</li></ul>

Figure 5-5. *Overall Overview of the Site Reliability Engineering Skillset*

When systems fail, we learn how a calm, structured approach to incident response transforms outages into learning opportunities, reducing MTTD and MTTR while fostering a culture of continuous improvement through blameless retrospectives.

Finally, we embraced the role of the SRE as a change agent, one who seeds best practices, advocates for automation-first thinking, and shifts organizational attitudes toward proactive reliability.

By mastering these skills, it is not just contributing to uptime but focusing on enabling a sustainable culture of operational excellence.

CHAPTER 6

SRE Transformation Exemplified on Two Deep Dives

In this chapter, we take a deeper look at two practical examples for rolling out SRE capabilities. First, we show how we tackle toil management—finding the right teams to pilot improvements and then scaling the approach step by step across the company. Second, we explain how to mature SLO practices over time and communicate them effectively to business partners to ensure alignment and buy-in.

Each capability follows a simple four-step rollout approach:

1. Identify and select pilot teams

2. Run controlled pilots and gather insights

3. Refine and expand the rollout to other teams

4. Embed the capability into daily operations and governance

Throughout this journey, it helps to use proven tools to enable success, for example, Jira for managing work items, Prometheus and Grafana for observability, and Nobl9 for tracking SLOs, while recognizing that the stack may vary.

From Manual Malaise to Automated Magic—The Imperative of Toil Management

At the very core of every SRE transformation in a technology organization, a silent battle rages. On one side, the relentless drive to innovate, to build, to create the future. On the other, the insidious creep of repetitive, manual work—the digital quicksand we call **toil**.

© Florian Hoeppner, Francesco Sbaraglia 2025
F. Hoeppner and F. Sbaraglia, *Mastering Site Reliability Engineering in Enterprise*,
https://doi.org/10.1007/979-8-8688-1448-8_6

In the old world, during the lean IT movement, it was called waste. This is the invisible anchor dragging teams down, sapping energy, and diverting focus from what truly propels a business forward. But on the other side, we experienced that some people in companies like this kind of repetitive manual work; if we start with automation and optimization, we experience pushback. Sometimes practitioners feel like they are losing their jobs when we start rolling out toil management.

But what exactly constitutes this "toil"? It's any operational task that Google defined as follows:

- **Manual:** It demands human hands and eyes for execution, often when a machine could perform it more reliably.

- **Repetitive:** It's not a one-off, but a recurring chore that gnaws away at valuable time, sprint after sprint.

- **Automatable:** Its underlying logic is predictable enough that software could handle it, often faster and with fewer errors.

- **Tactical and Reactive:** It's driven by immediate needs and alerts, rather than strategic, proactive design.

- **Devoid of Enduring Value:** Once completed, the system is no better off; no lasting improvement has been made. It's the digital equivalent of Sisyphus rolling his boulder, only for it to tumble back down.

At the beginning of our transformation, we need to put the definition in the words of the practitioners from our company. Imagine this: your brightest engineers manually provisioning servers each time a traffic spike looms, meticulously running diagnostic scripts (Example: Listing 6-1) after every deployment, or painstakingly entering identical data into a spreadsheet. It *feels* like a productive activity. But in reality, it's often just spinning wheels. This kind of work doesn't scale with your ambitions, it doesn't energize your talent, and it certainly doesn't inspire the breakthroughs that define market leaders. If your systems are getting more users, you also need more people to maintain and operate them.

Listing 6-1. Example Bash script to check and scale Kubernetes deployment

```bash
#!/bin/bash

# ---- Config ----
NAMESPACE="default"
DEPLOYMENT="checkout-service"
DESIRED_REPLICAS=3

# ---- Get current replica count ----
CURRENT_REPLICAS=$(kubectl get deployment "$DEPLOYMENT" -n "$NAMESPACE" -o
jsonpath='{.spec.replicas}')

echo "Current replicas for $DEPLOYMENT: $CURRENT_REPLICAS"

# ---- Compare and scale if needed ----
if [ "$CURRENT_REPLICAS" -lt "$DESIRED_REPLICAS" ]; then
  echo "Scaling $DEPLOYMENT to $DESIRED_REPLICAS replicas..."
  kubectl scale deployment "$DEPLOYMENT" --replicas="$DESIRED_REPLICAS" -n
  "$NAMESPACE"
  echo "Scaling initiated."
else
  echo "No scaling needed. Replica count is sufficient."
fi
```

When toil becomes the majority of work, the consequences for our teams are negative:

- **Increasing Costs:** Every manually repeated task is inefficient, resulting in a double-spending on time and resources.

- **Eroding Quality:** Human intervention, especially under pressure, is a gateway for errors, inconsistencies, and vulnerabilities. And when rework is required, this is mostly under pressure, and this is additional effort.

- **Compromised Security:** Critical security processes executed by hand are invitations for mistakes, oversights, or deviations from best practice.

- **The Human Toll—Burnout:** Talented engineers, hired for their creativity and problem-solving skills, become frustrated, disengaged, and eventually seek opportunities where their skills are better utilized.

Many practitioners working in global tech organizations, leaders in their industry, will find themselves grappling with these very symptoms. Across their technology organization, spanning often hundreds of engineers, the same refrains echoed: chronic alert fatigue, laborious manual deployments, time-consuming repetitive diagnostics, repeating incidents, checking log files each morning, and the mind-numbing duplication of data entry. The weight of this collective toil can be overwhelming. However, instead of succumbing to it as an unavoidable cost of doing business, some companies chose a different path. They resolved to systematically dismantle it by launching a structured, enterprise-wide toil management rollout. Their journey, detailed in these pages, serves not just as a case study. Still, we want to give here a practical blueprint for any organization ready to transform its operational landscape. Often, we start with toil management in our SRE transformation because it frees up time for the following other capabilities.

The Rollout—Step 1: Showing the Invisible— Making Toil Visible

The ancient wisdom holds true: you cannot fix what you cannot see. Toil, often a shadow activity, performed with grim acceptance that this must be done that way or so ingrained it becomes invisible, is the first thing that we need to enlighten. The initial, foundational step in our ambitious journey is to quantify this hidden monster on productivity. We can choose two lightweight, yet effective, mechanisms to achieve this. Our key is to not produce toil by managing toil:

1. **Ticket-Based Logging: Capturing Toil in the Workflow**
   The first mechanism should be integrated with existing habits. Every time an engineer or operations team member performs a task they recognize as toil, they are encouraged to tag their ticket as toil, or if they work without tickets, they should open a simple ticket. This wasn't meant to be an arduous process but a quick capture, including:

- A concise **description of the task** (e.g., "Manually restarted trading gateway service X").

- An estimate of the **time spent**.

  - A **tag describing the category of toil** (e.g., #toil_deployment, #toil_serverrestart).

Listing 6-2. Jira ticket export example with toil management tag

Title:

Automate Service Restart for CheckoutService Deployment

Description:

Currently, engineers manually restart the CheckoutService deployment when CPU usage exceeds 90%, which happens 2–3 times per week. This manual task is repetitive, interruptive, and adds no long-term value.
Acceptance Criteria:

- Create a script to monitor CPU usage and trigger an automatic restart when needed.

- Add monitoring and alerting to validate that the automation works as expected.

- Document the automation in the team runbook.

Labels:

toil-management automation reliability

Custom Field - Toil Reduction:

Toil Hours Saved: ~3 engineer hours/week

Priority:

Medium

Assignee:

SRE Team

Epic/Parent:

Reduce Manual Toil for Production Ops

This approach can best be embedded into toil tracking within the tools engineers already use—systems like Jira or, for operations people, ServiceNow. It minimized disruption, transforming routine ticketing into a powerful data collection engine. Suddenly, individual acts of toil began to aggregate, showing a picture of where the collective time was truly going, making toil transparent. The problem we are facing is that many teams do not track their work in any ticket system. Also, we realized that some are hesitating to write down the time they have spent. The second option is tracking that problem, but it does not lead to scientific results. The positive side of this approach is that we can scale it across the enterprise and aggregate the results if we want to approach some enterprise-wide automation.

2. **Sprint-Based Surveys and Discussions: Surfacing Latent Toil**
   Complementing real-time logging, a firm can introduce structured discussions about toil into its agile cadences. During sprint planning or, more commonly, in retrospectives, teams were prompted to reflect on:

   - What manual, repetitive tasks did they encounter during the sprint?

   - How frequently did these tasks occur?

   - What was the perceived time sink or frustration level associated with them?

   - What options do you see for automating it?

These regular, conversational check-ins proved invaluable. They helped validate the data trickling in from tickets and, crucially, surfaced "invisible toil"—those tasks so profoundly embedded in muscle memory that individuals had stopped consciously registering them as burdensome or even automatable. For this approach, we can use a survey tool, maybe anonymously. This will help teams to open up. On the downside, this approach will deliver vague results, and it can be challenging to aggregate results from teams at a domain or company level.

Why This Initial Step Is Pivotal:

It's tempting for technically minded organizations to leapfrog directly to automation solutions. However, automating without a clear understanding of the problem landscape often leads to misdirected effort and suboptimal outcomes. By first meticulously measuring and visualizing toil, the firm achieves several critical objectives:

- **Data-Driven Justification:** They constructed an undeniable, data-backed business case for change. Abstract frustrations were translated into quantifiable hours, which in turn translated to cost and opportunity loss. It helps to get the funding, set the priority, and keep the program going. We can start sharing the success stories of the hours of manual work we automated. Example Estimated Hours Saved = Σ (Task Frequency × Time per Task).

- **Building Credibility and Trust:** This transparent approach demonstrated to engineers that leadership was not only aware of their operational burdens but was also genuinely committed to alleviating them. Conversely, leadership gained a clear, unfiltered view into the daily realities faced by their teams. This shared understanding became the bedrock for the efforts that followed.

- **Toil Threshold:** Measurement puts us in the position to balance toil against other work. We can set a threshold on the team level, and when the threshold is reached, automation is guaranteed. Otherwise, leadership will always prefer to work on new innovative features instead of putting effort into automation.

- **Toil Aggregation from Team to Firm Level:** By tagging toil, we can find common patterns between teams. In some cases, it may not make sense for single team to pursue automation, or restrictions may hinder progress. At the firm level, however, there may be opportunities to invest in new automation tools, adjust policies, or develop shared programs that eliminate manual effort across the organization.

With toil now visible but measurable and transparent, and by reporting it to the business and leadership and having an argument to invest time in remediation, the stage is set for a more targeted attack.

Step 2: From Data to Diagnosis—Categorizing and Connecting the Dots

With a growing repository of identified toil, the next crucial phase is to move beyond mere quantification to deep understanding. It is not enough to know *that* toil existed; the firm and the teams needed to dissect *what* it truly represented and *where* its roots lay. This involved a systematic process of categorization (Example: Figure 6-1) for each logged item or survey-surfaced task:

The Anatomy of Toil—Key Categorization Dimensions:

- **Source System/Area:** Where is this toil originating? (e.g., legacy trading platforms, specific application monitoring, gaps in CI/CD pipelines, fragmentation due to mergers and acquisitions).

- **Type of Toil:** What is the nature of the manual work? (e.g., manual intervention post-deployment, repetitive data entry, reactive troubleshooting for recurring alerts, lack of self-service capabilities for users).

- **Frequency and Effort:** How often does this task demand attention, and how much time does it consume per instance? (e.g., daily, weekly, per-release; minutes, hours).

- **Impact and Consequence:** What is the broader business or operational effect of this toil? (e.g., delays critical market data delivery, increases incident resolution time, leads to customer-facing errors, causes regulatory reporting rework, impacts developer productivity).

Field	Example Value	Description
`toil.source`	CI/CD	Where the toil originates (e.g., CI/CD, Infra, Monitoring)
`toil.type`	manual deploy	The nature of the toil task (e.g., manual restart, repetitive config)
`toil.owner`	Team Alpha	Which team or squad owns the service/workload
`toil.freq`	5 per week	Frequency of occurrence
`toil.est_time`	15 mins	Estimated time per occurrence
`toil.status`	Identified / In Progress / Automated	Toil reduction status for reporting

Figure 6-1. *Toil Tagging Taxonomy Example*

This disciplined categorization acted like a prism, refracting the raw light of toil data (Example: Figure 6-2) into a spectrum of actionable insights. As teams meticulously tagged and analyzed their toil, profound patterns began to emerge for the teams and the enterprise. This can be, for example:

- **Onboarding Inefficiencies:** Inconsistent or overly manual onboarding processes were consistently generating a high volume of access requests and troubleshooting tickets for new systems.

- **Diagnostic Bottlenecks:** The reliance on manual diagnostic procedures during production outages can be a significant contributor to prolonged service restoration times.

- **Post-Merger Duplication:** The integration (or lack thereof) of systems following mergers and acquisitions can result in numerous parallel platforms requiring duplicated manual effort for data synchronization and reporting.

- **Alert Storms:** Poorly configured monitoring systems can trigger floods of unactionable alerts, leading to significant manual investigation and alert fatigue.

Toil Task	toil.source	toil.type	toil.freq	toil.est_time	toil.status
Manual deploy to staging	CI/CD	manual deploy	5 per week	15 mins	Identified
Service restart by script	Operations	manual restart	3 per week	10 mins	In Progress
Re-run failed data job	Data Pipeline	manual data re-run	2 per week	30 mins	Automated

Figure 6-2. *Example Toil Categorization Table*

The true power of this step is unleashed when data from dozens of teams can be aggregated and visualized. We can have sophisticated dashboards that provide views that could be sliced and diced by team, business domain, technology stack, and across the entire enterprise. But for the start, a query in Jira and a simple spreadsheet can be enough. A holistic perspective enabled leadership and teams to

- **Prioritize Strategically:** Identify common, high-impact toil patterns that warrant centrally funded and developed solutions (e.g., a unified identity management system).

- **Empower Local Action:** Pinpoint localized issues that individual teams could effectively automate or re-engineer within their sprints. Give the arguments for adding toil reduction tasks to the next sprint.

- **Justify Targeted Investment:** Build compelling cases for acquiring new automation tools, redesigning cumbersome processes, or investing in upskilling programs focused on specific automation technologies.

This phase is definitely transformative. It shifted the narrative around toil from one of individual frustration and anecdotal complaints to a shared, systemic understanding. Toil is no longer perceived as a personal inconvenience or a sign of individual failing but as an organizational challenge with identifiable causes and, crucially, solvable problems. Everyone now has a more precise map and also a shared language with the tags to navigate the path toward reduction.

Step 3: The Toil Takedown—Solve, Sustain, and Celebrate

Identifying and understanding toil are critical prerequisites, but the actual transformation lies in its systematic reduction and the establishment of an environment where it's continuously challenged. Toil management, the firm recognized, isn't a finite project with an end date; it's an ongoing operational capability, woven into the very fabric of how teams work. Toil management must become part of the DNA because with each change, toil is coming again and again. Someone small in pockets, but it will add up and become bigger and bigger. This third step focused on embedding this ethos through a multi-pronged approach:

1. **Empowered Teams: Owning the Day-to-Day Battle**

   The frontline in the war against toil was, and always will be, the individual engineering and operations teams. They were empowered and expected to treat toil with the same rigor they applied to feature development or incident management:

   - **Continuous Logging:** Toil tickets weren't an occasional task but must be logged consistently throughout each sprint, as manual, repetitive work was encountered.

   - **Retrospective Review:** Teams established their toil thresholds (e.g., "no more than 15% of sprint capacity spent on automatable tasks"). These thresholds were reviewed in sprint retrospectives, prompting discussions on what was causing breaches and how to address them.

- **Prioritized Automation:** Based on retrospective insights, tasks to automate specific toil items were explicitly prioritized and pulled into subsequent sprint backlogs, treated as valuable engineering work.

- **Daily Reinforcement:** Even daily stand-ups became a micro-forum for toil awareness. A quick mention like, "I had to restart the trade reconciliation queue again yesterday manually, logged it as toil, and we should look at automating that script," kept the issue top-of-mind without derailing the meeting. This lightweight, persistent process ensured toil reduction became an integral part of the team's rhythm and the company's DNA.

2. **Enterprise Enablement: Providing the Systemic Support**

   While teams drive the daily efforts, the central SRE or platform engineering function played a crucial role as an enabler and accelerator:

   - **A Common Platform and Taxonomy:** Standardized tooling (e.g., Jira, ServiceNow) is configured with predefined fields and tags for consistent toil logging and reporting across the organization.

   - **Centralized Dashboards and Analytics:** The central team maintained enterprise-wide dashboards, offering insights into toil trends and reduction progress and identifying cross-cutting concerns that might benefit from shared solutions.

   - **The "Toil Academy"—Knowledge and Resources:** A repository of best practices, cheat sheets, short video tutorials, and scheduled "office hours" with automation experts can be established to guide teams.

   - **Tool Access and Training Provision:** The organization invested in and provided access to, along with relevant training for, key automation technologies identified as high-leverage (e.g., Infrastructure-as-Code tools like Terraform, CI/CD orchestrators like Jenkins, RPA platforms like Blue Prism, or scripting languages like Python). On a central platform, automation scripts, solutions, and champions can be shared with other teams. This platform with solutions helps to share between teams.

- **Transparent Reporting Engine:** A system should be established to share automation progress and toil reduction metrics with all stakeholders on a regular (e.g., monthly) basis, fostering accountability and showcasing successes. This transparency ensures that business and leadership see the value, motivating continued investment in automation.

3. **Cultivating a Culture of Automation: Motivation Makes It Work**

   Tools and processes are essential, but sustained change requires a cultural shift. A firm should cleverly implement several "cultural levers" to ensure that toil reduction is not seen as just another mandate but as a desirable and recognized contribution:

   - **Nudging Toward Automation:** Postmortems for incidents or significant operational events now routinely include the question, "What aspects of the detection, diagnosis, or remediation of this issue were manual and could be automated to prevent recurrence or speed recovery next time?"

   - **Gamification and Healthy Competition:** Reward teams with digital badges for hitting toil reduction goals or for particularly innovative automation solutions.

   - **Public Recognition and Storytelling:** "Automation Heroes"— individuals or teams who made significant strides in eliminating toil—should be celebrated in town halls, newsletters, and internal communication channels. Their stories inspired others.

   - **Leadership Transparency and Commitment:** Dashboards showcasing overall toil reduction progress were regularly reviewed by senior leadership, signaling its strategic importance.

   - **Dedicated Innovation Time—Hackathons:** Quarterly hackathons, with specific tracks and prizes for toil automation projects, unleashed creativity and generated practical solutions in a focused, high-energy environment.

Through these concerted efforts, automation transcended being a mere task on a checklist. It evolved into an aspiration, a mark of engineering excellence, and a tangible way for teams to reclaim their time and refocus on more engaging, value-added work.

Step 4: Institutionalizing Excellence—Toil Management As an Enduring Capability

The true triumph of this rollout is not just the immediate reduction in manual effort but the profound shift that occurred *after* the initial push. Toil management transcended being a special project and should become an ingrained, institutional muscle—a living, breathing capability continuously flexing and strengthening across the organization. This durable transformation is normally anchored by five core foundations:

The Five Foundations of Sustainable Toil Management:

1. **A Compelling Vision:** The narrative is not simply "eliminate toil"; it is reframed to "transform toil into opportunity." Teams are not just being asked to stop doing grunt work; they are being invited to liberate their time and intellect for higher-impact, more innovative endeavors. This aspirational framing was key to winning their hearts and minds.

2. **Continuous Measurement and Improvement:** The adage "what gets measured gets managed" remained central. Two primary North Star metrics should be diligently tracked:

   - **Adoption Rate:** The percentage of teams actively measuring, categorizing, and working to reduce their toil. Example Adoption Rate = (Active Teams Logging Toil / Total Engineering Teams) × 100

   - **Outcome Metric:** The total number of engineering hours saved (or reclaimed) through automation initiatives, often translated into dollar savings or capacity unlocked for innovation. (e.g., 50 hours/week x 52 teams = ~2,600 hours reclaimed).

We cannot underestimate the power of these two metrics. When toil programs stop after a short time, it's primarily because of low adoption or an uncommunicated outcome. Mainly, when we show that a combination of increasing adoption generates improved outcomes for the teams, the users, and the firm, our toil initiatives rollout gains speed. We can measure adoption by tracking training effort, the number of teams tracking toil, showing detected toil effort, teams with toil thresholds, and finally, teams within their threshold.

3. **A Dedicated Platform and Ownership:** A clear **Capability Owner** (often within a central SRE, platform engineering, or operational excellence team) should be designated, supported by a small core team. Their responsibilities included

   - Maintaining and evolving the tools and integrations for toil tracking (Jira, ServiceNow, etc.).

   - Governing the tagging taxonomy to ensure consistency and relevance.

   - Curating and enhancing the enterprise dashboards and reporting mechanisms.

   - Championing the overall toil management strategy.

   - Creating a portal with toil reduction solutions

   - Providing awareness, training for toil management, maturity steps, and tracking of adoption and outcome

   - Providing a forum for sharing success stories, for example, leadership calls where the team can present how they track toil and solutions they use for the reduction of toil.

4. **A Thriving Community of Practice:** To foster peer-to-peer learning and maintain momentum, **Toil Champions** should be identified or volunteered within every central region, business line, or significant engineering department. These champions

- Acted as local evangelists and first points of contact for toil-related queries.

- Participated in regular (e.g., monthly) calls to share best practices, discuss common challenges, and showcase notable automation wins from their respective areas. This created a vibrant, self-sustaining ecosystem of knowledge exchange.

5. **Embedded Upskilling and Support:** Recognizing that automation requires specific skills, the organization committed to ongoing learning opportunities:

- Engineers need access to a curated library of training materials, workshops, and pathways to certification in relevant automation tooling and scripting languages.

- "Automation Clinics," or dedicated office hours (e.g., every "First Friday"), should be established, where engineers could bring their specific toil challenges and receive guidance from experts or peers on how to approach automation.

A Concluding Reflection: Beyond the Balance Sheet

This systematic rollout can do far more than just exercise hours of manual labor from the organization's daily operations. It forged stronger bonds of trust and collaboration between engineering teams and leadership, as both parties worked toward a common, visible goal. It demonstrably frees up highly skilled engineers from the drudgery of repetitive tasks, allowing them to redirect their talents toward innovation, feature development, and strategic problem-solving. The firm's operations became more resilient, less prone to human error, and quicker to adapt.

Perhaps most importantly, this initiative must cultivate a culture that profoundly respects its engineers' and operations' time and intellectual energy. In an era where talent retention is paramount, budgets are perpetually scrutinized, and the specter of burnout looms large, the most intelligent and enduring investment a technology organization can make is strikingly simple: proactively identify and systematically reduce the work that shouldn't exist at all. By turning toil into a catalyst for improvement, a firm not only enhanced its efficiency but also enriched its engineering culture, proving that the path from manual malaise to automated magic is achievable with vision, structure, and commitment.

In the second half of the chapter, we want to show a deep dive into the rollout of the SLO/error budget capability.

Rolling Out SLOs and Error Budgets: Weaving Reliability into the Fabric of Your Organization

Introduction: The Dual Mandate of Modern Engineering

In the relentless competition between companies and their digital products, downtime is never an option. Uptime, responsiveness, and a seamless user experience are no extras; they are the fundament of user expectation and key to building user trust. It's the invisible contract upon which trust between users and firms is built. Yet, simultaneously, the clarion call for innovation is louder than ever. Product teams are under immense pressure to ship new features, iterate rapidly, and conquer new markets. This inherent tension—the walk between unwavering reliability and velocity—is where many well-intentioned organizations falter; often, you can read it in the news days later.

Enter service level objectives (SLOs) and their indispensable counterpart, error budgets. These are not mere metrics; they are core tenets of site reliability engineering (SRE), offering a pragmatic, data-driven framework to navigate the complex trade-offs between risk and innovation. They transform abstract desires for "stability" and "speed" into a concrete, shared language that aligns engineering, product, and business stakeholders.

Aspect	Definition	Example (Login Latency)
SLI (Service Level Indicator)	A measurable metric that indicates the current performance of a system or feature.	*95th percentile login request latency.*
SLO (Service Level Objective)	The target goal or threshold for the SLI over a specific time period.	*95% of login requests must respond in under 500ms over the last 30 days.*
SLA (Service Level Agreement)	A formal, contractual commitment between provider and customer, often with penalties if the SLO is missed.	*If average login latency exceeds 500ms for two consecutive weeks, customer receives service credits.*

Figure 6-3. *Examples SLI, SLO, and SLA for Login Request Latency*

Following, we want to give you another field guide, forged from collective experience, to successfully implement SLOs and error budgets within your tech organization. We'll navigate a strategic, phased rollout—strategic alignment, pilot and learn, and general enablement—in addition, we will share illustrative stories from the trenches, actionable help, and hard-won tactics you can adapt and deploy. Our goal is to help you embed reliability not as an afterthought, but as an intrinsic part of your engineering and operations practitioners' DNA.

Step 1: Strategic Alignment—Laying the Cornerstone for Cultural Change

Embarking on an SLO adoption journey without first securing strategic alignment will not bring you anywhere. Mainly with this capability, because it's nothing where you just need one team, here we change the way teams work together—business, engineers, production support, release management, and more. This initial phase is about building a shared understanding and a compelling case for why reliability, quantified and managed, is critical to the organization's success.

The Art of the Narrative: Making Reliability Everyone's Business

Before a single SLO is defined, before any dashboard is configured, you must win the attention of your stakeholders and get them to understand what you are going to improve. Your first task is to craft a compelling narrative that changes departmental silos and positions reliability as a collective imperative. This isn't just a technical exercise; it's organizational change management.

- **Illustrative Story: The Digital Bank's Weekend Woes:** Consider "FIRST FinBank," a rapidly growing digital bank. Their engineering team, celebrated for its agile prowess, launched a groundbreaking new payments feature leveraging a social media platform late on a Friday. Technically, the system remained "up"—their legacy monitoring showed 99.95% availability. Yet, over the weekend, social media erupted. Users reported excruciating latency in transaction processing, with some transactions timing out entirely. Not only some, but more and more were getting worse. While the development team had already moved on to the next sprint, the operations team was fighting the firestorm of incident calls, and customer support lines were choked. Databases with payment data were crashing, and data was getting lost. The reputational damage was palpable, and the cost of remediation, both in engineering hours and lost customer trust, was steep. Had FIRST FinBank adopted SLOs (e.g., "<400 ms latency at 95th percentile") and Error Budgets, the narrative could have been different. An SLO tied to transaction latency would have triggered alerts early. The corresponding Error Budget depletion would have signaled an unacceptable regression, potentially pausing the problematic rollout or triggering an immediate, focused rollback *before* widespread customer impact.

To build your narrative, we need to frame reliability as

- **A Shared Responsibility:** Emphasize that reliability isn't just "Ops' problem" or "SRE's job." It's a collaborative effort involving product (defining user expectations), engineering (building resilient systems), and operations (maintaining and responding).

- **A Competitive Differentiator:** In a crowded market, trust is currency. A reliable service retains users, attracts new ones, and builds brand loyalty. Conversely, unreliability erodes trust and drives users to competitors.

- **A Quantifiable Promise:** SLOs transform vague aspirations like "the system should be stable" into a measurable, actionable commitment to your users, both internal and external.

- **Reliability Is a Business Decision:** SLOs and Error Budget consumption are making reliability transparent. The final decision on how reliable a system should be lies with the business, because we need to balance it with cost (see Figure 6-4).

Optimal Reliability Is a Business Decision

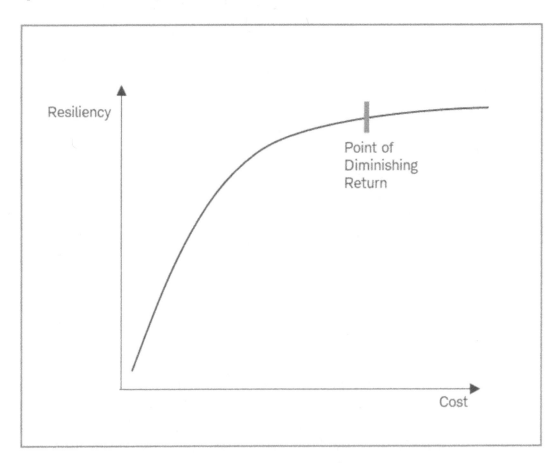

Figure 6-4. *Example Optimal Reliability Trade-off Graph*

Balance speed, cost, and reliability trade-offs: align the reliability objectives with the business context.

Demystifying the Trinity: SLI, SLO, and Error Budget

With the "why" established, let's clarify the "what." These three concepts are inextricably linked:

- **Service Level Indicator (SLI):** This is a direct, quantitative measure of some aspect of the service's performance from the user's perspective. An SLI isn't just any metric; it's a carefully chosen proxy for user happiness. How a user experiences and feels your application. Examples include request latency (e.g., percentage of requests served in under 300 ms), availability (e.g., proportion of successful requests), data freshness, or system throughput.

- **Service Level Objective (SLO):** This is the target value or range for an SLI over a specified period. It's a precise numerical goal (e.g., 99.9% of login requests should complete successfully within 500 ms over a rolling 28-day window). The SLO is a statement of intent, a commitment to a certain level of service quality.

- **Error Budget:** This is the critical, empowering consequence of setting an SLO. It represents the permissible amount of unreliability—the tolerance for events that cause the SLI to miss its SLO target. It is the mathematical inverse of your SLO.

  Formula: Error Budget=100%−SLO Percentage

  Example: If your SLO for API availability is 99.9% per month, your Error Budget is 0.1%. For a 30-day month (approximately 43,200 minutes), this translates to roughly 43 minutes and 12 seconds of permissible "badness" (e.g., downtime, excessive errors, or unacceptable latency) before the SLO is breached. This budget isn't for "slacking off"; it's a calculated risk allowance that empowers teams to innovate, deploy new features, and perform maintenance, knowing precisely how much "unhappiness" they can introduce before violating their user promise.

Marrying SLOs with Business Imperatives: The Tiering Strategy

Acknowledge the reality: not all services carry the same weight in the grand scheme of your business. Attempting to apply a one-size-fits-all SLO (like the mythical "five nines" for everything) is not only impractical but also a misallocation of resources. A more pragmatic approach involves tiering services based on their criticality and business impact.

Tier	Example Service	Suggested SLO (Availability)	Justification
Tier 0	Core authentication API, payment gateway	99.99–99.999%	Absolutely mission-critical
Tier 1	Main application features, order processing	99.9–99.95%	High business impact
Tier 2	Reporting dashboard, batch analytics	99.5–99.9%	Important for business
Tier 3	Internal experimentation platform, Dev tools	99–98%	Low direct impact

This tiering directly informs where engineering effort, incident response urgency, and investment in resilience should be focused. It provides a clear rationale when discussing trade-offs with product and business leaders. And we expect most professional companies to have set this up already.

Step 2: Pilot and Learn—Validating the Model in the Real World

The temptation to mandate SLOs organization-wide can be strong, especially after achieving initial strategic buy-in. Resist it. A broad, simultaneous rollout is fraught with peril and likely to overwhelm teams. They will tell you that they don't have time for that. It was all running well the last few years; no need to do anything. So instead, opt for a carefully curated pilot program. This is where theory meets practice, allowing you to learn, iterate, and build credibility.

Finding the Pioneers: Selecting Your Pilot Teams

The success of your pilot program hinges significantly on the teams you choose. Ideal candidates exhibit:

- **A Well-Understood Production Service:** The service should have clear user journeys and a relatively stable operational history. Avoid services already mired in constant firefighting.

- **Existing Observability Foundations:** The team should already have some level of logging, metrics collection, and ideally, tracing in place. You need data to define SLIs.

- **A Mature (or Maturing) Incident Management Process:** The team should be accustomed to responding to incidents, conducting postmortems, and learning from failures.

- **Enthusiasm and Willingness:** Look for teams eager to improve and open to new methodologies. A skeptical but willing team can also be a powerful advocate if won over.

Pro Pilot Mix Tip: Consider selecting two distinct types of teams for your pilot:

1. A team managing a critical, customer-facing application (e.g., the shopping cart API).

2. A team responsible for a key internal platform or tool (e.g., the CI/CD pipeline). This blend provides diverse learning opportunities, showcasing SLO applicability across different service types and user bases.

3. You also want to have more than one pilot team. Normally, I start with three pilot teams, because one or two might get sidetracked or something might not be working right away.

Defining the First Crucial SLOs: Start Simple, Iterate Often

Avoid the "analysis paralysis" of trying to define every conceivable SLO for your pilot services. Begin with one or two high-impact SLIs per service. These should ideally reflect the most critical aspects of the user experience.

Example for an E-commerce Product Page Service:

- **SLI 1 (Availability):** Percentage of successful HTTP GET requests (2xx status codes) for product detail pages.

    - **SLO 1:** 99.9% over a rolling 30-day window.

    - **Error Budget 1:** 0.1% (approx. 43 minutes of unacceptably high error rates per month).

- **SLI 2 (Latency):** Percentage of product detail page requests served in under 400 ms (at the 95th percentile).

    - **SLO 2:** 99% of requests (P95) served < 400 ms over a rolling 30-day window.

    - **Error Budget 2:** 1% of requests can exceed this latency threshold before the SLO is breached.

Crucially, verify that your existing observability stack (e.g., Prometheus, Grafana, Datadog, New Relic, OpenTelemetry collectors) can accurately capture the raw data needed for these SLIs. If not, enhancing instrumentation becomes a prerequisite. Check out the OpenSLO concept and validate if you want to consider a tool as a central point for all your SLOs, like Nobl9.

From Noise to Signal: Implementing Error Budget-Based Alerting

One of the most profound shifts SLOs enable is the evolution of alerting strategy. Move away from reactive, often noisy, threshold-based alerts (e.g., "CPU utilization > 90% for 5 minutes!") toward user-centric, SLO-driven alerts based on error budget consumption.

- **The Core Principle:** Alert when the rate of error budget consumption threatens to breach the SLO within its compliance period.

- *Example Alert Logic:*

    - "Warning: Service X has consumed 50% of its monthly error budget for latency in the first 7 days. Projected to breach SLO if current trend continues."

- "Critical: Service Y has consumed 10% of its monthly error budget for availability in the last hour due to a spike in 5xx errors. Immediate investigation required."

- Consider setting alerts for various burn rates:

  - **Fast Burn:** When a significant portion of the budget (e.g., 2%, 5%, 10%) is consumed in a short window (e.g., 1 hour, 6 hours). This catches sudden, severe regressions.

  - **Budget Depletion Thresholds:** When total consumption reaches milestones like 25%, 50%, 75%, and 90% of the budget within the compliance period. This tracks slower, "death by a thousand cuts" degradations.

This approach dramatically reduces alert fatigue by focusing attention on what directly impacts users and the service's reliability commitments. It prioritizes incidents based on actual or imminent SLO violation.

Weaving SLOs into the Team's Agile Rhythm: Integrating with Workflows

SLOs and error budgets are not just for dashboards; they must become integral to the team's daily, weekly, and sprint cadence.

Workflow	SLO Integration Example
Sprint planning	If the error budget is low (<25% remaining), proactively allocate engineering capacity for reliability improvements or bug fixes. De-prioritize risky new features.
Daily stand-ups	Briefly discuss current error budget status and burn rate trends, especially if concerning.
Release gates	Use error budget status as a key criterion for go/no-go decisions. A depleted budget might warrant pausing a release.
Postmortems	Every incident investigation should include an analysis of SLO impact and error budget consumption. Did our SLOs detect this? If not, why?
Capacity planning	Use historical SLO performance and error budget consumption to forecast future infrastructure and resource needs.

During the pilot, work closely with teams to identify natural integration points and refine these workflows.

Step 3: General Enablement—Scaling the Practice Static and Constant

With successful pilots under your belt, armed with valuable lessons and internal champions, you're ready to scale the adoption of SLOs across the wider organization. This phase is about systematizing what you've learned and empowering all teams to embrace this new way of operating.

Cultivating a Movement: Training, Evangelism, and Storytelling

Scaling requires more than a mandate; it demands education, and we need to inspire:

- **Tailored Enablement Sessions:** Develop training modules specific to different roles:

  - **Developers:** How SLOs inform resilient design, coding best practices, and the impact of their changes on user experience. How to use error budgets for safe experimentation.

  - **Product Managers:** How to articulate user needs as measurable SLOs, how to balance feature velocity with reliability targets, and how error budgets provide a framework for risk-taking.

  - **SREs/Ops Engineers:** Deep dives into SLI selection, SLO mathematics, advanced alerting strategies, and automating SLO reporting.

  - **Executives and Leadership:** How SLOs link to business objectives, improve customer satisfaction, optimize resource allocation, and foster a culture of accountability.

- **Amplify Success Stories:** Leverage the experiences of your pilot teams. Their testimonials, challenges overcome, and tangible benefits (e.g., "We reduced P1 incidents by 30% after implementing SLOs") are potent fuel for broader adoption. Internal tech talks, newsletters, and brown-bag lunches are great forums.

Paving the Road: Standardizing Definitions, Tooling, and Dashboards

Consistency is key to scaling effectively. Strive for standardization to reduce cognitive load and ensure everyone is speaking the same language.

- **SLO Templates and Libraries:** Create predefined templates for common SLI types (e.g., availability, latency, freshness) using your organization's preferred configuration language (YAML, JSON, Terraform modules). This accelerates SLO definition for new services.

- **Playbooks for Setting Thresholds:** Provide guidance on how to determine appropriate SLO targets based on service tiers, historical performance, and user expectations.

- **Standardized SLO Dashboards:** Design common dashboard layouts (e.g., in Grafana, Datadog) that all teams can adopt. This provides a consistent view of

  - Current SLI value

  - SLO target

  - Remaining Error Budget (percentage and absolute time/count)

  - Error Budget burn rate (e.g., 1x = on track, >1x = burning too fast)

  - Time to SLO breach at current burn rate

  - Historical SLO compliance

Example of a standardized dashboard panel (conceptual):

```
→ Service: User Authentication API
SLO: 99.95% Availability (SLI: Successful Logins / Total Login Attempts)
Window: Rolling 28 days
Current SLI: 99.97%
Error Budget Remaining: 85% (approx. 18m 12s for this period)
Burn Rate (last 24h): 0.75x (Healthy)
Time Since Last SLO Breach: 52 days
```

Closing the Loop: Linking SLOs to Governance and Incident Response

To give SLOs real teeth, integrate them into critical governance processes.

- **Error Budgets as Policy:** Formalize the concept of the "Error Budget Policy." For instance:

  - If Error Budget > 50%: Proceed with planned releases and experiments with standard caution.

  - If 25% < Error Budget <= 50%: Exercise increased caution. Consider deferring non-critical or high-risk changes. Prioritize reliability work.

  - If Error Budget <= 25% (or breached): Freeze non-emergency feature releases. All available engineering effort focuses on restoring reliability and replenishing the budget. A formal review may be required to resume normal development.

- **Incident Management Enrichment:** Ensure SLO status and error budget consumption are prominent in:

  - **Incident Timelines:** How quickly was the budget burning during the incident?

  - **On-Call Dashboards:** Is the on-call engineer aware of services close to breaching their SLOs?

- **Release Readiness Reviews:** Is the service healthy enough (from an SLO perspective) to absorb the risk of a new deployment?

- **Postmortem Reports:** Was the SLO breached? If so, what was the impact, and how can future breaches be prevented?

Step 4: Sustaining and Evolving a Culture of Reliability

Implementing SLOs is not a one-time project; it's the beginning of an ongoing journey. Sustaining this practice requires diligence, continuous improvement, and a commitment to keeping reliability at the forefront.

The Rhythm of Review: Institutionalizing Continuous Improvement

SLOs are not static artifacts to be defined and forgotten. They must live and breathe with your services and user expectations.

- **Regular SLO Reviews:** Establish cadences for reviewing SLO performance:

  - **Team Level (e.g., Bi-weekly or Monthly):** In retrospectives or dedicated reliability meetings, teams should discuss their SLO adherence, error budget burn, any near-misses, and identify actions to improve.

  - **Organizational Level (e.g., Quarterly):** Review overall reliability trends, assess if SLOs are still appropriate, and identify systemic issues or opportunities for improvement. Recalibrate SLOs based on evolving business needs, user feedback, and observed system capabilities. Has a Tier 2 service become Tier 1 due to increased reliance?

Fostering Accountability Without Blame: The Psychological Safety Net

A critical element of a successful SLO culture is psychological safety. Error budget depletion or SLO breaches should trigger investigation and learning, not finger-pointing or punishment.

- **Focus on System, Not Individuals:** Emphasize "What happened, and how can we prevent it?" rather than "Who broke it?"

- **Learning Opportunities:** Treat SLO breaches as invaluable signals that the system (or process) needs attention. They are triggers for engineering improvements, not indictments of individuals or teams.

- **Reward Proactive Reliability Behavior:** Recognize and celebrate teams that proactively manage their error budgets, pause risky releases when necessary, or contribute significantly to improving service resilience—even if it means delaying a feature. This reinforces that reliability is valued alongside velocity.

The Symbiosis of Reliability and Innovation: Enabling Velocity with Confidence

One of the most powerful, yet often misunderstood, aspects of a mature SLO/Error Budget model is its ability to *accelerate* innovation, not hinder it. By providing clear guardrails and objective data, SLOs empower teams to make informed decisions about risk.

- **Clarity on Risk:** Teams understand precisely how much "room for error" they have. A healthy error budget can greenlight more experimental features or faster deployment cadences. A dwindling budget signals a need for caution and focus on stability.

- **Data-Driven Decisions:** Instead of relying on gut feelings about whether a release is "too risky," teams can use error budget consumption as a data point.

Case in Point: The "FIRST SwiftRetail" Transformation. FIRST SwiftRetail, an e-commerce platform, was struggling with the classic reliability versus velocity dilemma. After a comprehensive SLO rollout, guided by the principles outlined here, they witnessed a remarkable shift. Over six months, they reduced the frequency of critical production incidents by 40%. Simultaneously, their average deployment frequency for their core services *increased* by 25%. How? Teams used their error budgets intelligently. They felt more confident pushing updates when budgets were healthy, and they knew precisely when to apply the brakes, temporarily freezing changes to focus on stability when budgets ran low. This data-driven approach replaced fear-based decision-making with calculated confidence.

Charting the Journey: SLO Maturity Steps and Reporting to Leadership

The rollout and adoption of SLOs across an organization is a significant transformation, not an overnight switch. To effectively manage this journey, motivate teams, and clearly communicate progress and outcomes to leadership, it's invaluable to define SLO maturity steps. We provide a framework as a roadmap, allow for celebration of milestones, and underpin the narrative of increasing organizational reliability and efficiency. Reporting against these steps is not just an update; it's a strategic tool to reinforce commitment, secure resources, and drive the cultural shift.

Example of Defining SLO Maturity Levels:

While every organization's path is unique, a general model for SLO maturity can be adapted:

- Level 0: Unaware/Ad Hoc

    - **Characteristics:** No formal SLOs exist. Reliability is typically reactive, driven by outages and customer complaints. Monitoring is often system-centric (CPU, memory) rather than user-centric. "Best effort" is the prevailing approach.

    - **Reporting Focus (to Leadership):** All teams target in the organization.

- Level 1: Initial/Piloting

  - **Characteristics:** The concept of SLOs has been introduced. One or two pilot teams are defining their first SLIs and SLOs for well-understood services. Basic dashboards are being built. Focus is on learning, experimentation, and identifying champions.

  - **Reporting Focus** (to **Leadership**):

    - **Adoption:** Number of pilot teams engaged, key services selected for piloting.

    - **Progress:** Initial SLIs/SLOs defined for pilot services.

- Level 2: Defined/Expanding

  - **Characteristics:** SLOs are defined for a growing number of critical services. Standardized templates and tooling for SLO tracking and dashboarding are emerging. Error budget concepts are understood, and alerting based on budget burn is being implemented. Teams are beginning to discuss SLOs in sprint planning and postmortems.

  - **Reporting Focus (to Leadership):**

    - **Adoption:** Percentage of Tier 0/1 services with defined SLOs. Number of teams actively using SLOs.

    - **Outcomes (Early Indicators):**

      - Anecdotal improvements in incident detection for services with SLOs.

      - Reduction in alert noise for pilot/early adopter teams.

      - Qualitative feedback from teams on clarity and focus.

- Level 3: Managed/Integrated

  - **Characteristics:** SLOs cover the majority of business-critical and user-facing services. Error budget policies are actively used to gate releases and influence engineering priorities. SLO reviews are a regular part of team and inter-team cadences. Reliability work driven by SLO data is explicitly planned.

- **Reporting Focus (to Leadership):**

  - **Adoption:** Percentage of all relevant services (Tier 0, 1, 2) with SLOs.

  - **Outcomes (Measurable Impact):**

    - Demonstrable reduction in SLO-breaching incidents for covered services.

    - Improvement in key SLI metrics (e.g., sustained higher availability, lower latency).

    - Correlation between error budget health and deployment success rates.

    - Quantifiable reduction in incident response times or severity for SLO-covered services.

- Level 4: Optimized/Embedded

  - **Characteristics:** SLOs are deeply ingrained in the engineering culture. They are proactively reviewed, refined, and used not just for operational stability but for strategic capacity planning and even informing business decisions. Error budgets are used to strategically balance risk and innovation across the organization. There's a continuous feedback loop where insights from SLOs drive architectural improvements and preventative measures.

  - **Reporting Focus (to Leadership):**

    - **Adoption:** Near-universal SLO coverage for impactful services.

    - **Outcomes (Strategic Impact):**

      - Clear data linking reliability improvements (via SLOs) to business KPIs (e.g., customer satisfaction, retention, reduced operational costs).

      - Evidence of faster, yet safer, innovation cycles enabled by error budget management.

- Use of SLO data for long-term forecasting and resource allocation.

- Stories of how a strong reliability posture became a competitive advantage.

Reporting is important for motivation, to get buy-in in the upcoming phases, and to have a smooth transformation. Regularly communicating this journey to leadership, and indeed to all stakeholders, is key:

1. **Visibility and Validation:** It makes the often-invisible work of reliability tangible and showcases the positive impact of the SLO initiative. This validates the effort of the teams involved.

2. **Sustained Sponsorship:** Leadership sees concrete progress and ROI (in terms of stability, efficiency, or even customer satisfaction), making them more likely to continue sponsoring the initiative with resources and political capital.

3. **Motivation and Healthy Competition:** Sharing successes (e.g., "Team X improved their critical service's uptime SLO from 99.9% to 99.95% last quarter") can motivate other teams and foster a spirit of continuous improvement.

4. **Accountability and Course Correction**: Reporting against maturity levels keeps the transformation on track. If progress stalls, it becomes a visible discussion point, allowing for intervention and support.

5. **Reinforcing Cultural Change:** Each report that highlights SLO-driven decisions and positive outcomes reinforces the message that reliability is a priority and that this new way of working is valued.

By framing the SLO rollout as a maturity journey and consistently reporting on both adoption and outcomes, you transform it from a technical project into a strategic business initiative. This level of communication is critical for inspiring teams, aligning stakeholders, and ensuring that the pursuit of reliability becomes an enduring part of your organization's DNA.

Summary

SLOs are the heartbeat of engineering excellence. Rolling out service level objectives and Error Budgets is far more than a technical initiative; it's a profound cultural transformation. It's about fundamentally altering how your teams perceive system health, engage with user experience, and achieve a sustainable, balanced velocity.

By embracing a phased approach, from securing strategic alignment and proving value through pilots to scaling the practice thoughtfully across the organization, you cultivate not just more reliable services but also more empowered and data-informed engineering teams.

Let your SLOs and Error Budgets transcend the confines of mere reports and dashboards. Let them become the living, breathing heartbeat of your engineering culture—a constant, guiding rhythm that drives both resilience and innovation, ensuring your services not only run but thrive.

## CHAPTER 7

Scaling SRE: Business Benefits and Value Drivers

In this chapter, SRE will not just be a technical concept but a living, breathing piece, a hardcore and active element of our corporate culture. We will be able to pitch and convince business stakeholders to invest in SRE by showing clear evidence of its impact. As we transition into this pivotal chapter, we will delve deeper into the core benefits and value drivers of site reliability engineering (SRE) within the enterprise landscape. Previously, we outlined the structure and significance of a central enablement team, a crucial component of the center of excellence (CoE). This team plays an essential role in leading strategic initiatives, establishing effective governance, and securing quick wins that showcase measurable value, for example, reducing toil hours, improving SLO adherence, increasing end-to-end observability of critical business processes, and lowering incident recovery time, thereby cultivating momentum within the organization.

A key takeaway from our earlier discussions is that the success of our SRE implementation hinges on our ability to market our achievements successfully and swiftly rebound from setbacks, using those experiences to enhance our approach continuously. In this chapter, we will explore the compelling reasons why businesses should allocate their time and resources to SRE initiatives and develop a robust return on investment (ROI) framework tailored for business stakeholders.

We will address how to identify key enterprise-wide value drivers that resonate with our organizational goals. Moreover, we will construct a comprehensive business case for SRE, detailing how these initiatives can lead to significant, quantifiable benefits. To effectively measure the success of SRE practices, we will discuss essential KPIs and metrics, supported by widely adopted tools and platforms such as **OpenSLO, Nobl9, Dynatrace, or Grafana**, that provide crucial insights into performance and effectiveness. Finally, we will confront the organizational challenges and necessary changes that often arise when scaling SRE practices, ensuring we are well-prepared to navigate these complexities.

© Florian Hoeppner, Francesco Sbaraglia 2025
F. Hoeppner and F. Sbaraglia, *Mastering Site Reliability Engineering in Enterprise*,
https://doi.org/10.1007/979-8-8688-1448-8_7

By the end of this chapter, SRE will be more than just a technical initiative; it will be recognized as a dynamic and integral part of our corporate culture. We aim not only to educate but also to equip stakeholders with the understanding needed to champion investments in SRE as a transformative journey.

This chapter will cover the following main topics:

- **Identifying Key Enterprise-Wide Value Drivers**

- **ROI and Business Case for SRE**

- **Measuring SRE Success with KPIs and Metrics**

- **Addressing Organizational Impediments and Changes at Scale**

- **Identifying Key Enterprise-Wide Value Drivers**

Identifying Key Enterprise-Wide Value Drivers

Identifying key enterprise-wide value drivers involves recognizing the fundamental factors that contribute to an organization's overall value and success, ensuring alignment with strategic goals, and maximizing operational efficiency.

Let's now look at one concrete case study from the retail industry. Why did we select Retail? It is an evolving industry with many new best-of-breed technologies supporting fast business growth, such as Metaverse or edge computing. These technologies impact our daily lives as consumers, and minor improvements in this industry are typically well perceived by the end-user, so by us directly.When a large retailer, we can call the company here **ShopCo**, began their digital transformation journey, they quickly ran into an **equilibrium** between getting new features fast on the market and stability issues. Their e-commerce platform saw multiple outages, one after the other, first of all by increasing the target audience, but later even by the typical seasonal events, like Black Friday, or just the launch of new trendy articles. The driver for the increased traffic is just the run of multiple marketing campaigns; this has cost millions in lost revenue. The management approved a new SRE initiative to improve resilience and get back the trust of the business department.

The SRE team jumped into action. They were close to the issues and had the right skills and capacity. We added three outside SRE consultants to the team, given the importance and for advice on the unknown ground. They defined critical customer-impacting failure scenarios, increased latency, duplicate orders, and payment service errors. Then, we conducted a game day, and I remember it well. It was a hot Friday, and it was getting late. We injected real-world faults into production. Some issues we experienced, some from our external consultants, knowing similar cases. Initially, these experiments overwhelmed teams, and we had some minor complaints. We wanted to get faster. We had to slow down.

After six months, ShopCo's management requested complex numbers to renew funding. Despite preventing incidents, the SRE team, which now had a dedicated Chaos Engineering team, needed to democratize the practice, demonstrate direct financial impact, and enable teams to self-service new chaos experiments. While they tracked experiment findings, they didn't consistently link them to quantifiable business value. The team regrouped to identify relevant metrics they already had good quality data for, such as **P95 latency deviations**, **error injection success rates**, and **system recovery time after failure injection** (Example: Table 7-1).

Table 7-1. *Chaos Experiment Report Example, Key Findings, and Potential Financial Impact*

Chaos Experiment	Key Finding	Potential Financial Impact
Injected network latency	P95 latency spiked 250 ms under load	Identified API bottleneck ➤ reduced cart drop-offs ➤ estimated +$X revenue retention
Database failover test	Recovery time exceeded the 5 min threshold	Informed DB cluster redesign ➤ reduced downtime risk ➤ avoided SLA penalties
CPU stress test on checkout	Error rate rose by 3%	Led to auto-scaling tuning ➤ improved conversion rates during peak sales

The SRE/Chaos Engineering Lead regrouped their team to identify relevant metrics from which they had good quality data. They found two key areas:

- **Revenue Loss**: By tracking e-commerce platform availability and transaction volume, they quantified direct revenue impact. Improved availability resulted in a million incremental monthly sales.

- **Operational Efficiency**: Decreased incident response effort due to proactive activities saved a couple of thousand dollars yearly in engineer time. Fewer Sev 1 production incidents also built confidence around faster feature releases.

Armed with these metrics tied to value, the SRE team presented a compelling business case to management. The numbers spoke for themselves—over a couple of million in annual bottom-line savings. Chaos Engineering received extended funding to expand tests and drive resilience as ShopCo's competitive advantage.

This case shows why it is critical to start with the end in mind to demonstrate value. The team reacted late, but luckily, we were still able to get the numbers they needed. They saved the funding and suddenly gained a reputation. While setting up a Chaos Engineering team is crucial, quantifying financial and efficiency impact is what convinces executives. With rigorous measurement discipline, chaos engineering can transform system resilience and unlock business performance.

Another three critical points of this case study are

- **Self-Service**: For chaos experiments to be feasible, self-service must be enabled for all teams (e.g., LitmusChaos, ChaosMesh, Azure Chaos Studio, Steadybit, etc.). It is vital to keep track of significant experiments, especially disruptive ones, and be careful if this is the very first experiment. Recently, self-service using templates has enabled teams to incorporate chaos engineering into every CI/CD pipeline.

- **Automation**: The first one or two experiments were hands-on, such as scripting in Bash to inject latency (Example: Listing 7-1), creating IPtables firewall rules to block traffic, or just running kubectl delete pod. Later, the adoption of a chaos engineering platform and tool saved precious time for the SRE teams.

Listing 7-1. Example script introducing network delay of 200 ms to a Linux host

```bash
#!/bin/bash
# Target network interface
IFACE="eth0"
# Add 200ms of network latency
sudo tc qdisc add dev $IFACE root netem delay 200ms
echo "START 200ms latency added on $IFACE"

# Keep it for 60 seconds
sleep 60
```

```
# Remove the added latency to restore normal conditions
sudo tc qdisc del dev $IFACE root netem
echo "STOP Latency removed from $IFACE"
```

- **Enablement**: Continuous enablement sessions with multiple stakeholders and teams provided good foundational knowledge across numerous teams and individuals. SRE ambassadors helped script and use chaos experimentation more deeply at the core of the microservices and platform.

Table 7-2. *Example Checklist of 6-Day Enablement Chaos Engineering*

Day	Focus Topic	Example Activities and Takeaways
Day 1	*Introduction and foundations*	What is SRE and Chaos Engineering? Understand failure modes and real-world examples. Why Chaos Engineering to increase resilience?
Day 2	*Platform-specific fault injection*	Hands-on: Inject latency or CPU stress on staging; intro to tools like Chaos Mesh or Steadybit.
Day 3	*Observability and steady state hypotheses*	Define SLIs/SLOs; practice writing a hypothesis for a chaos experimentation. Create and use dashboards to monitor impact. Learn how to document any condition's change of the system.
Day 4	*Application-specific fault injection*	Hands-on: custom application fault injection on staging; develop new advanced application use cases in tools like Chaos Mesh or Steadybit.
Day 5	*Designing safe experiments*	Define blast radius; plan rollback; practice using safeguards like litmus probes, infra and application smoke tests, and observability signals.
Day 6	*Run game days and share learnings*	Run a full game day; practice incident communications; do a blameless retro; update runbooks.

Having explored a lighthouse case study of a retail company successfully implementing Chaos Engineering, we've gained valuable insights into the practical benefits and strategies for enhancing system reliability and operational efficiency. These real-world examples have highlighted the importance of key performance indicators (KPIs) in tracking and measuring the impact of Chaos Engineering initiatives.

By leveraging these KPIs, organizations can develop a strategic approach to improve their processes continuously. Last but not least, it is crucial to create assets and know-how around the three critical points of automation, self-service, and enablement.

Now, as we transition to the next topic, we will focus on reducing the toil (manual tasks) associated with Chaos Experimentation. Site reliability engineering (SRE) defines toil as manual and repetitive work with low value that should be automated. Toil can significantly impact efficiency. By minimizing toil, we can streamline chaos experimentation, making it more effective and sustainable. Let's explore the methods and tools available to reduce toil and enhance the overall efficiency of our Chaos Engineering practices.

ROI and Business Case for SRE

Some topics sound good on paper and theoretically, but they will never be measured, and it is challenging to track them and link them back to our work.

Other topics are hidden; we do some great work, but no one will notice. This happens the whole time because everything runs smoothly and well. It is the life of someone responsible for production support. The magic in the background sees none. When I set up some advanced monitoring, we observed an upcoming issue, and I fixed the bug in a night job before the servers stood still. One notices, maybe my manager, when I charge overtime; maybe my supervisor tells me I cost too much because I charged overtime. The point I want to make is when we get real, we must take incidents into account that have yet to surface. That is, the point of SRE is to be prepared to answer some tricky questions from business and advertise those incidents as a learning point for the team that will be ready for the near future.

Our target as "Manager" or "Ambassador" of SRE must be to lead discussions with facts and numbers. We measure everything related to the SRE movement. But in reality, for starters, only some numbers count. We must have one clear metric, the amount of money we saved, to show the tangible impact of SRE. Ultimately, it all leads back to the number of incidents, which we translate into dollars. We have two options to calculate this: one is based on the number of incidents and the total downtime of an application; the second is to track the number of issues addressed proactively. I prefer doing both calculations; they will lead you to a corridor where the truth lies. To make these estimates credible, you can reference industry benchmarks, for example, *Gartner *(https://www.the20.com/blog/the-cost-of-it-downtime/) and Atlassian (https://www.atlassian.com/incident-management/kpis/cost-of-downtime) have found that average downtime costs can range from $5,600 to $9,000 per minute, depending on the business. This gives your financial impact calculation a strong, defensible baseline.*

Option 1: Calculate Based on the Number of Incidents

Assessing the costs associated with downtime over time is essential. As downtime is not always transparent, is often not clearly measured, and we have part downtimes, I prefer to calculate incidents. Downtime can be for some users only, for some functions, and for a region. Figuring out how to calculate the details for all applications in scope can be challenging. Because of that, we use incident tickets and measure the following for a given time (typically monthly):

- The number of incidents

- Time to solve an incident, MTTR = Total Downtime (in minutes)/ Number of incidents

- Severity (Business Criticality) of an Incident (high, medium, low)

- Application Criticality from the Incident (high, medium, low)

We measure this over time and validate it with historical data and against similar applications. Typically, we see identical solution times when the application has the same technology and maturity level. We multiply the number of incidents by the resolution time and an average salary (depending on the location of the support) to get the amount for the support. Based on the decrease, we can calculate how much we saved.

This approach has some downsides:

- We cannot argue that SRE caused the positive impact. Other circumstances might have reduced the effort.

- Some companies and teams do not have high-quality incident data because people do not open incidents in some cases, and the resolution time may lead to a false conclusion (ticket open and closed times are the same, tickets were closed after years, etc.).

Recommendation:

We should start using the incident ticket data. The data will improve only when we use it and start reporting.

Option 2: Calculate Based on Addressed Issues

In our SRE tasks, we produce findings and actions to improve the system. These findings result in actions that lead to a change in the application, the system, the team, the infrastructure, or even vendor contracts. This will give us two numbers per application:

- **Identified Issues**: These are the findings after our experiments

- **Addressed Issues**: When we have an improvement, for example, a fix in the system, we can close the issue

Table 7-3. *Example Incident Risk and Resolution Dashboard*

Issue ID	Estimated Outage Probability	Estimated Revenue Risk	Resolution Effort (Hours)	Notes
INC-1012	High (70%)	$50,000/hour	8	Legacy DB failover needs an upgrade
INC-1045	Medium (40%)	$20,000/hour	3	API latency spikes under peak load
INC-1098	Low (10%)	$5,000/hour	1	Minor config drift—quick fix

We want to enrich this data with two factors:

- **Likelihood**: The probability that this issue could have happened, for example, that this could have led to an outage

- **Assumed Impact**: The impacts relate to the business criticality of the application; this can include reputation loss, and users are not able to work

With all those numbers, we can again calculate the amount our work has saved our company. In this case, we can even calculate the time we invested to fix the bugs when our teams collected estimates and implementation efforts.

This approach has some downsides:

- We rely on our teams to track and validate the numbers. We might need to invest in a dashboard, etc. Other than incidents, this might be new data for a company.

- Most teams will need some guidance to estimate the likelihood and the accumulated impact, which can cause considerable variances in the numbers.

Recommendation:

The effort might be higher at the start to collect the numbers and hold teams accountable for discussing and collecting the numbers. In addition, it requires effort to have good data quality and to remind teams.

Measuring SRE Success with KPIs and Metrics

We noticed that some companies have monitoring systems in place and that the ITSM (IT Service Management) team tracks various indicators. As we explained in the first part, the link between business outcome and indicators, and in the second part of our recommendation, we understand that some teams must start "lean." They track some metrics and still need the initial capacity to build a new dashboard and align **Key Performance Indicators (KPIs)**.

The following are the KPIs, calculations, and business relevance that are primarily impacted by the SRE initiative.

Table 7-4. *Performance KPIs Impacted by SRE Implementation*

KPI	Definition	Unit	Impact	Business Value
Cost per incident	Average financial cost to resolve an incident	$	Down	Operational efficiency
System availability	Proposition of when the system can be used by the user when it's required	%	Up	Risk management
Change fail rate	Number of changes that require a roll-back, roll-forward, or a fix	%	Down	Operational efficiency
Incident volume	The sum of the number of incidents per application compared to the number for the last period	%	Down	Operational efficiency
Mean time to resolve	The average time to resolve an incident compared to the number of incidents in the last period	%	Down	Operational efficiency
Incidents per severity	The number of incidents per severity (low, medium, high) indicates the business's criticality.	%	Down	Operational efficiency

Here's the detailed explanation for each KPI from Table 7-4:

- **Cost per incident or cost per outages** measures the financial impact of each incident on the organization. This KPI includes direct costs such as labor, resources used in the resolution process, and any lost revenue or penalties incurred (Example: Figure 7-1). Indirect costs, such as the impact on customer satisfaction and potential reputational damage, may also be considered. Calculating the cost per incident helps organizations understand the financial implications of downtime and improve their incident management processes to minimize these costs. Every cent saved through better incident management can effectively auto-finance the SRE initiative (*e.g., average cost per incident from Ponemon Institute* https://www. vertiv.com/globalassets/documents/reports/2016-cost-of-data-center-outages-11-11_51190_1.pdf).

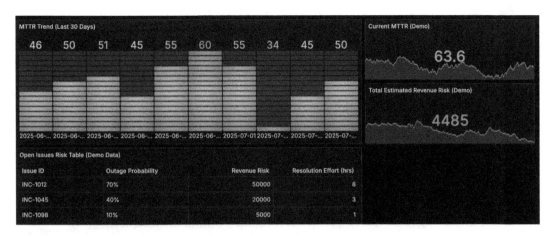

Figure 7-1. *Example SRE Dashboard Cost per Incident, Risk Table, and MTTR*

- **System availability** is a crucial KPI that reflects the reliability of a system. It refers to the percentage of time a system is operational and accessible to users. It is typically measured as a percentage over a specific period, such as monthly or annually. High system availability indicates reliable and consistent performance, while lower availability suggests frequent outages or downtime. This KPI is crucial for maintaining user trust and ensuring that services are delivered as expected.

- **Change failure rate** is an important KPI for assessing the stability of system changes. It measures the percentage of system changes that result in failures, incidents, or rollbacks. This KPI helps organizations evaluate the effectiveness of their change management processes. A high change failure rate may indicate inadequate testing or poor implementation practices, while a low rate suggests that changes are being successfully integrated without causing disruptions.

- **Incident volume** Incident volume provides insight into the overall frequency of issues within a system. It refers to the total number of incidents reported within a specific period, typically monthly or quarterly. This KPI helps identify trends or patterns that may need addressing. In large organizations with diverse teams and systems of varying scale, it's important to adjust incident counts for fair comparison. One effective approach is to standardize how incidents are tracked, for example, reporting incidents **per 1,000 API calls**, **per sprint**, or **per unit of transaction volume**. Normalizing the data in this way helps avoid misleading or anomalous trend lines that can occur when services handle vastly different amounts of traffic. This ensures the data reflects true reliability patterns, so teams can spot where to focus improvement efforts and direct resources where they'll have the most impact.

- **Mean time to resolve (MTTR)** is a critical KPI for evaluating the efficiency of incident management. It measures the average time taken to resolve incidents and restore the working functionality from the moment they are reported until they are fully resolved. This KPI is critical for understanding the efficiency of the incident management and automation processes in place. Lower MTTR indicates that issues are addressed and resolved quickly, minimizing downtime and its impact on the organization.

- **Incidents per severity** help organizations prioritize their response to different types of incidents. This KPI categorizes incidents based on their severity levels (e.g., critical, major, and minor) and tracks the number of incidents in each category over a specific period. It helps organizations understand the impact of incidents on their operations and prioritize responses based on severity. By analyzing incidents per severity, organizations can identify areas that need improvement and allocate resources to address the most critical issues first.

In this section, we've explored key performance indicators (KPIs) that help us measure improvements in system reliability, cost efficiency, team collaboration, and overall operational effectiveness. These KPIs are not just numbers; they help develop a strategy to track improvements brought by SRE at each stage of the capability implementation. With a solid understanding of these KPIs, we can now move forward to discuss how reducing SRE's toil can further enhance our efficiency and streamline our processes.

Addressing Organizational Impediments and Changes at Scale

In chapter three, we introduced and explained the core changes to the Enterprise Operating Model; in this section, we want to get more specific and discuss the general concept of a new Tech Operating Model that must help to navigate into the "New" and maximize the values of implementing SRE in each area:

- Modern tech strategies amplify dynamic iterations based on **Agile Principles**. SRE complements agile development, such as Scrum, LeSS (large-scale scrum framework), and SAFe (scaled agile framework). With SRE, teams can test system resilience early and in a structured way. The Agile movement embraces quick experiments with customers to gain insights; SRE extends this by adding learning from non-functional tests and delivering fast feedback to engineers on real-world faults. To embed this mindset, many teams now weave SRE practices directly into Agile ceremonies, for example, incorporating reliability goals into sprint planning, reviewing incident learnings and error budgets during retrospectives, and highlighting resilience trends during sprint reviews or demos. Additionally, modern SDLC deployment pipelines can integrate **fault injection** directly into CI/CD workflows, for example, by adding a custom **Chaos Engineering stage** in GitLab that runs Steadybit experiments or Chaos Mesh scenarios against staging environments. This ensures resilience is tested continuously alongside functional and security checks, giving teams real confidence before code reaches production.

- DevOps practices aim to increase deployment velocity. They are the same as Agile but with a higher focus on automation. SRE principles, with the spotlight on automating experiments and doing it in a highly transparent way, give the confidence to ship faster. Engineers understand the system's boundaries better despite the increasing system complexity.

- For the last few years, the anchor point underlying each tech strategy has been moving into the public cloud and gaining momentum in the adoption of **Cloud Native Principles**. As enterprises adopt cloud-native technologies, this results in a more dynamic infrastructure-as-a-service, which is described in scripts and outside of our direct control, hidden in a black box from a 3rd party distributed over the globe. In this constant change, our defined experiments (Example: Listings 7-2 and 7-3) are kept the same. We can rerun them and analyze the differences. The described SRE principles support our need to understand continuous change. SRE is the anchor for stability.

Listing 7-2. Example git repository structure

```
chaos-experiments/
├── checkout-service/
│   ├── latency-injection.yaml
│   ├── pod-delete.yaml
├── payment-service/
│   ├── cpu-stress.yaml
│   ├── network-partition.yaml
├── README.md
```

Listing 7-3. Example chaos mesh runbook network latency fault injection

```
apiVersion: chaos-mesh.org/v1alpha1
kind: NetworkChaos
metadata:
  name: checkout-latency
  namespace: chaos-testing
spec:
  action: delay
  mode: all
  selector:
    namespaces:
```

```
    - default
  labelSelectors:
    app: checkout-service
delay:
  latency: '200ms'
duration: '5m'
```

- Many architects decouple systems and build **microservices** and distributed systems, which increases system complexity. SRE simplifies the understanding of complex system behaviors. On a game day, our experiments give details on whether non-functional requirements are still met. When we inject real-world faults, we gain greater clarity into how end-user experience is impacted across services, uncovering latent issues before they create outages.

- As enterprises digitally transform, availability and stability become more important. Corporations rely more on software and technology to conduct business and reach customers. Everything is a digital application, an app nowadays. Applications, infrastructure, and services have become mission-critical, and outages that were once tolerable can now significantly impact revenue and reputation. Historically, many organizations operated in a reactive, incident-driven mode: wait for an incident, fix it, open a problem ticket, investigate, and make incremental improvements. SRE is different; it offers a proactive approach to avoid incidents before they happen. SREs don't wait for incidents to occur; instead, they spend most of their time anticipating potential failures and proactively adapting systems to prevent them. It's important to note that SRE brings clear benefits to both **brownfield** (**legacy**) and **greenfield** (**new**) environments. In brownfield systems, SRE practices help modernize reliability for existing, often complex infrastructure, for example, by adding observability layers and gradually automating repetitive manual fixes. In greenfield projects, SRE enables teams to build reliability from day one, designing with clear SLOs, error budget policies, and resilience patterns from the start. This distinction helps leaders justify ROI: brownfield improvements reduce legacy risks and

hidden costs, while greenfield investments prevent new issues and
scale with business growth.

- The last change modern tech strategies suggest is **autonomous
  teams**, which are built in a product and platform structure. In
  contemporary operating models, autonomous teams require
  new testing approaches. They are empowered to make their own
  decisions, and traditional end-to-end testing does not fit them.
  SRE maintains velocity while ensuring quality, providing teams
  with the confidence to move fast by constantly experimenting and
  understanding how systems behave under change. For example,
  many organizations see release velocity improve by 20–30% after
  adopting SRE practices, while still maintaining SLO compliance
  above 99.9%, proving that reliability and speed can scale together.

Traditionally, we look at benefits from the perspectives of **revenue increase** and **cost
reduction**. In addition, we want to set the spotlight on the people behind tech. Even if
this is not tangible, because when people are leaving the company and the team, this has
a cost impact, we need to set this as a highlight.

Revenue increase means attracting more customers, increasing sales volume,
raising prices, enhancing customer retention, increasing the frequency with which
customers use the service/product, or rolling out to more markets and increasing the
number of customers.

From the SRE perspective, we support **shortened time-to-market** for new features
and products by rapidly validating production readiness. For that, we leverage tooling
and automation. Increasing the availability of the application and reducing revenue
losses from outages is attracting more customers. This can be one of the critical points
when customers decide on a service. When we proactively find weaknesses, we gain
the trust of our teams and the customer. The customer experience will be smooth, with
fewer errors and better responsiveness, which will increase the joy of using the service.

Cost reduction means doing something more efficient. Automation of manual
tasks, reducing energy, improving the process steps and handovers, and finding
problems earlier in the process can help. With the insight gained in SRE, we can reduce
infrastructure costs by *right-sizing capacity* based on actual failure-tolerant limits. By
proactively finding weaknesses, we decrease the effort in incident response, primarily

if this would have led to an outage. The automation of our experiments reduces testing time compared to manual testing.

The last point is the **improvement of engineering life**. Working on the weekends, deployments late on Friday, and fixing incidents late at night are the main pain points. None would like to get pinged and paged by an alert for an incident and must fix a bug or application problem under time pressure. With SRE, we *increase the confidence* in deployments and let developers improve the velocity enabled by our automated testing. Being responsible with *increased ownership and empowerment* to ensure system resilience is possible with the transparency SRE gives. The cross-team collaboration gives a *sense of belonging*, letting us help other teams when we share our experiments and learnings. This clarity and the open discussion with teams working downstream or upstream of our application show the dependencies. Our engineers gain confidence by proactive fault injection; they prefer it over firefighting late at night. All of this leads to *higher job satisfaction* not only for those directly operating the system but for the entire team.

In the last section, we discussed linking business outcomes with revenue increases, cost reductions, and improvements in engineers' daily lives. Now, let's examine the actual benefits, using some examples from the industry.

Summary

This chapter concluded our exploration of SRE and its crucial role in enhancing both business and technological outcomes. We emphasized how SRE synergizes with agile software development, enabling teams to assess system resilience proactively through iterative testing. By quantifying the business advantages—like reduced incident response costs—we illustrated the potential for SRE to support and expand crucial initiatives.

Table 7-5. *Example Flow SRE Investment and Business Outcomes*

SRE Inputs	Metrics	Business Outcomes
Toil reduction efforts	MTTR, % automated tasks	More engineering time for innovation
SLO and Error Budgets	SLO compliance %, incidents avoided	Fewer SLA penalties, improved customer trust
Chaos experiments	Resilience test coverage, recovery time	Fewer surprise outages, higher system availability

This simple flow illustrates (Table 7-5) how SRE investments, like toil reduction, clear SLOs, and chaos engineering, feed measurable reliability metrics, which in turn drive tangible business outcomes such as lower downtime costs, increased deployment velocity, and greater customer confidence.

Moreover, we highlighted the value of automation in minimizing manual "toil," thus boosting operational efficiency and reallocating resources towards innovation. As we wrap up, it's clear that effectively measuring our enhancements is vital to demonstrating SRE's worth to stakeholders. In the upcoming chapter, we will delve into the tools available to help us automate workflows before, during, and after our experiments, further enhancing our efficiency.

## CHAPTER 8

Tools and Techniques for Scaling SRE

In the fast-paced world of site reliability engineering (SRE), tooling serves as the backbone for delivering consistent, efficient, and scalable operations. Yet with countless options available (e.g., Dynatrace, Splunk, Prometheus, Gremlin, LitmusChaos, Grafana, LaunchDarkly, Azure Chaos Studio, Steadybit, Harness, etc.), ranging from open-source to commercial solutions, selecting the right toolset can be overwhelming. This chapter dives into the **considerations and tradeoffs** every SRE team should weigh, details a structured approach for evaluating each tool's **Why, How, and What**, and culminates with critical takeaways in the form of **our learnings and summary**.

This chapter will cover the following main topics:

- **Assessing tooling requirements for enterprise SRE**

- **Chaos experiments, observability platforms, and AIOps**

- **Automation and orchestration techniques**

- **Evaluating trade-offs and integrations**

Assessing Tooling Requirements for Enterprise SRE

Once a tool is integrated, the likelihood of transitioning to another diminishes due to the considerable investment in time and resources associated with corporate procurement, which can span 6–10 months. Furthermore, as teams become accustomed to and proficient with a particular tool, the cost and effort required to switch tools increase, further reinforcing the initial choice. Therefore, the selection of the right SRE tool demands careful deliberation, ensuring it aligns with both our immediate and long-term needs.

© Florian Hoeppner, Francesco Sbaraglia 2025
F. Hoeppner and F. Sbaraglia, *Mastering Site Reliability Engineering in Enterprise*,
https://doi.org/10.1007/979-8-8688-1448-8_8

How to Select the Proper Tool for the Enterprise

We have crafted a versatile framework for selecting an SRE tool, which stands out not only for its specificity to our current challenge but also for its adaptability to other complex decision-making scenarios. This framework operates along eight dimensions, each representing a critical facet to consider when evaluating and choosing a tool that will best serve our organization's needs both now and in the future.

In the context of SRE, these eight dimensions will provide a structured approach to assess various tools. They allow us to compare their capabilities, scalability, integration ease, and the overall impact they might have on our systems and processes. The comprehensive nature of the framework ensures that we look beyond the immediate functionality of the tools and consider aspects such as the support ecosystem, community, learning curve, and the potential for the tool to evolve alongside our company's growth.

As we delve into the details of how to apply this framework, we will align each of the eight dimensions with the specific requirements of SRE, ensuring a thorough and nuanced selection process. This structured approach not only streamlines the decision-making process but also equips us with a method that can be replicated for other technological or strategic decisions we may face in the future.

Basically, when we select a tool for our enterprise, eight dimensions should be considered:

- **Strategic Alignment:** *Does this tool support the enterprise's long-term business goals and transformation roadmap? For example, selecting a cloud-native tool to support a multi-cloud migration strategy.*

- **Functional Adequacy:** *Does it have the required features and capabilities to meet our specific use cases? For example, advanced application custom injection and Kubernetes microservice-oriented versus standard infrastructure and performance injection only*

- **Technical Fit:** *Will it integrate well with our existing architecture, tech stack, and operating model? For example, on-prem version only versus cloud-first version, ensuring a new CI/CD tool works with our Kubernetes clusters and existing pipelines.*

- **Financial Fit:** *Is the total cost (licensing, implementation, and operations) justifiable and sustainable? For example, comparing licensing models for SaaS versus on-prem versions.*

- **Security and Compliance:** *Does the tool meet our regulatory, data privacy, and security requirements? For example, selecting a data analytics tool that complies with GDPR or IRAP certification.*

- **Usability:** *Is it intuitive and easy for our teams to adopt, with minimal training overhead? For example, a tool with self-service reporting to reduce training needs.*

- **Integration:** *Can it connect seamlessly with our other systems, data sources, and workflows? For example, a tool with out-of-the-box connectors for AWS, Azure, and GCP.*

- **Scalability:** *Can the tool grow with our business and handle increasing loads without performance issues? For example, a solution that supports horizontal scaling for global expansion.*

Scalability: *Can the Tool Grow with Our Business and Handle Increasing Loads Without Performance Issues?*

In our selection process, we want to understand if the tool and the vendor fit our company and strategic alignment. It should complement existing tools, development and operation processes, and vendor strategy. When I have a good existing relationship with a vendor, it might be easier to include a tool from a similar vendor. Or we have an open-source-first strategy, so we want to rate such a tool higher. In addition, you want your tools to keep on improving, and you receive regular feature and security updates. When you have a good relationship with your vendor or if you use an open-source tool, you may be able to influence the future product roadmap.

From the functional perspective, this is not only about the technical functionality. This included aspects for reporting and organizational alignment. When developing our tech strategy for high resiliency and stability, we need to have a dashboard and reports for different management levels. This is not required when we expect only the team level to work with the data. APIs can give us a structured way to extend functionality, but this comes at a price.

The technical fit looks to the integration of our new tool in the existing landscape. Does it meet the requirements for device support, maybe mobile support for dashboards and reports? Audit requirements, traceability, version control? Cloud support or multi-cloud support can be critical.

The financial fit included the commitment the company wanted to make for the long term. Different license models and consumption models are on the market. The tool should give admins an overview of adoption and usage. The financials can spike when many teams are testing the new tool, but maybe they will not use it for long. We require an overview of adoption and delivered outcomes. Otherwise, license models can fall short, and we pay a high price for limited usage and a high price for limited usage. We need to consider that modern cloud and SaaS pricing models, such as per-container or per-service billing, can lead to unexpected cost spikes if usage grows faster than anticipated. To avoid budget surprises, it's good practice to run a proof-of-concept (PoC) billing dashboard during the evaluation phase. This helps teams see how real usage patterns might affect costs before committing to a full rollout, ensuring the solution remains financially sustainable as you scale.

The security and compliance dimension takes the longest to validate in a corporate setting. Security architecture must be rigorously reviewed and tested by dedicated teams before any final decision is made, so early involvement is essential. It's also important to consider specific regulatory requirements and industry standards—for example, SOC 2, ISO 27001, or GDPR—as part of your tool selection process. Keep in mind that some open-source tools may lack certifications or built-in compliance features, which can add complexity to your risk assessments and controls.

We see a high value in having a managed test phase with multiple stakeholders to select the right tool. For that we select the teams who want to test, create scorecards, and agree on requirements and weighting. Our experience shows that the higher the number of teams involved in the testing, the better the acceptance after the rollout.

The following list will help to select the tool with the right features and technical aspects. This varies for each capability. Here we give the example for Chaos Engineering:

- Attack types

- Can stop attacks in progress

- Can attack containers

- Can attack serverless

- Test case integration

- Relevant API support (REST, CLI, YAML integration, etc.)

- Enterprise support

- GUI/UI/Web Portal

- Open source

- SAAS

- Development language (most are written in GO)

- Dependence on Linux Net Emulator?

- Agents and Demon model

In the next step, we want to evaluate potential tools. This starts with research and finally comparing different tools available, as an example for Chaos Engineering: *Steadybit*, *Gremlin*, *Verica*, *ChaoSlingr*, and *Deciduous*. We assess the capabilities of each tool to meet the organization's requirements, including features mentioned before, such as fault injection, monitoring, and analysis. We normally end up with two scorecards. One will be handed to the testers in our pilot and testing phase, and the second one will be used by the evaluation team and filled by multiple stakeholders to evaluate the vendor's reputation, support, and roadmap for the tool.

Now we are coming to the phase where we pilot and test multiple tools. We have narrowed down our shortlist to three or four candidates, with a technical team assembled from multiple business units and a clear technical scorecard in place. Next, we conduct a pilot or proof-of-concept deployment of each chaos engineering tool in a non-production environment. The testers inject faults and failures to observe how the system reacts and identify potential vulnerabilities. Typical pilot objectives include tracking the percentage of faults injected successfully, measuring the time taken to detect and assess the impact, and evaluating the effectiveness of rollback or self-healing mechanisms. Testers analyze the data from these experiments to determine each tool's effectiveness in meeting the organization's needs. Finally, all testers submit their scorecards with their ratings for each tool to support an informed final decision.

The last phase is where we implement, maintain, and drive the adoption of the tool. If the pilot is successful, we usually work with the tool provider to deploy the chaos engineering solution across the organization. Integrate the tool into the organization's existing processes, such as CI/CD pipelines, monitoring, and incident response.

The crucial part is that we think holistically to make the investment worthwhile for our company. A tool that is not used or not used in the right way is not worth its money. We want to train our people, update our processes and guidelines, and implement a governance board with regular reporting on adoption and outcome metrics.

Observability Platforms and AIOps by Michele Dodic

Today, more and more deployments and automated tasks are based on AIOps solutions. This section will depict how chaos experimentation can be used to test the resiliency of such solutions by applying the continuous cycle of hypothesis and experimentation. Furthermore, a use case will be presented, which combines SRE, Chaos Engineering, AIOps, and observability tooling and best practices, to demonstrate how SRE can leverage AIOps to fine-tune chaos experiments.

Essential Concepts: Chaos Experiments, AIOps, and Observability

First of all, let us define some key concepts:

- **Chaos Experiment**: It is a controlled process that introduces proactively disruptive events to a system to prove its resilience and ability to sustain disruptions under adverse conditions. Many organizations align these experiments with an established chaos maturity model, such as the OpenChaos framework or Gremlin's Chaos Maturity Model, to benchmark their progress, from running simple, isolated tests to embedding continuous chaos into CI/CD pipelines as their confidence and system resilience mature.

- **AIOps**: In short, Artificial Intelligence for IT Operations is the application of machine learning and data science to IT operations tasks and problems. It involves the use of big data, analytics, and artificial intelligence techniques to automate the identification and resolution of common IT issues. The goals of AIOps are to improve IT operations efficiency, predict and prevent potential problems before they impact users, and enable faster resolution of incidents. This is achieved by analyzing large volumes of historical operational data and identifying patterns that IT operators (humans) might not be able to detect, thereby facilitating more proactive and intelligent decision-making. It was originally defined by Gartner (here is a link to a deep dive into the topic: `https://www.gartner.com/en/information-technology/glossary/aiops-artificial-intelligence-operations`).

- **Observability**: A measure of how well internal states of a system can be inferred from knowledge of its external outputs.

- **Golden Triangle of Observability**: Refers to metrics, logs, and traces (some vendors now emphasize "events" as a fourth observability pillar).

- **Golden Signals**: Refer to latency, traffic, errors, and saturation.

AIOps can be a powerful solution, and if correctly implemented, it is able to predict any type of service degradation within the target system before its occurrence. However, one will have very limited possibilities if the target system is not first made observable. The complexity of modern systems and their interwoven dependencies make error detection and root cause analysis increasingly challenging tasks. Therefore, the goal is to leverage observability in order to gain visibility into the target system and all of its dependencies.

Upon reaching this milestone, the next objective is to implement AIOps. This involves leveraging the golden signals (facilitated through observability) and analyzing historical data from numerous system components, with the goal to not only identify the root cause of an existing issue but to predict potential failures before they occur. This is how, with AIOps, the paradigm shifts from reactive to predictive.

Continuous Cycle of Hypothesis and Experimentation

In the realm of Chaos Engineering, another pertinent concept involves the continuous cycle of hypotheses and experimentation, as depicted in Figure 8-1.

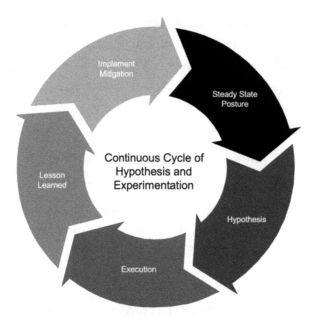

Figure 8-1. *Continuous Cycle of Hypothesis and Experimentation*

Let us define the steps:

1. **The Steady State Posture:** It indicates the current, as-is state of the target system before the initiation of the chaos experiment.

2. **A Hypothesis:** It is made based on the steady-state posture, which predicts the effect that the chaos experiment will induce once applied to the target system. A hypothesis is usually defined like this: the target system is resilient to the disruption of a specific set of services. In other words, the assumption is that the target system is sufficiently resilient, robust, and reliable to withstand the disruption of those services and will nevertheless continue to operate in a functional state.

3. **Continuous Verification:** It refers to the step in which the hypothesis is put to the test by actually applying the chaos experiment to the target system and observing the results.

4. **Lessons Learned**: Once the chaos experiment has been completed (and the services referred to in the hypothesis have been impacted), it is necessary to conduct a thorough investigation of the consequences that the experiment had on

the target system. This can be done by analyzing and measuring the health and stability of the overall system, post-disruption. For instance, if the chaos experiment was targeting a specific database, a post-investigation would require identifying whether the corresponding replicas have been successfully spinned up and whether this action has avoided any impact on the overall system health. If the replicas were spinned up successfully and the system health has been preserved, this means that the target system was able to withstand the chaos experiment. In the case in which, for instance, a replica did not spin up successfully, and a set of services of the overall system have been negatively affected, it means that the target system was unable to withstand the experiment, and therefore, it is necessary to identify the reason behind this occurrence and summarize the lessons learned.

5. **Implement Mitigation:** If we consider the last-mentioned case in which the target system fails to withstand the disruption of the chaos experiment, it is necessary to apply the lessons learned from the previous step and mitigate the failure. In simple terms, it is required to patch up the weak points (in this case, the failing replica). The motivation behind this step is to apply improvements to the failing components or services in order to ensure that the overall health of the target system remains unaffected once the same chaos experiment is run again in the next cycle.

After the implementation of mitigation is completed, a subsequent phase of the cycle can commence. During this phase, the new steady state resulting from the previously applied mitigation is taken into account, serving as the basis for formulating a new hypothesis. If the hypothesis from the previous cycle failed, the new hypothesis could be an identical one, as the goal is to test the resiliency of the improved target system (if we refer to the earlier example, these would be the enhanced replicas). Conversely, if the prior cycle proved successful, the subsequent cycle could be utilized to assess the resilience of a different component or service. Alternatively, one might opt to continue testing the same one, but with more assertive chaos experiments. As evident, this constitutes an ongoing process aimed at progressively enhancing the resilience of the target system.

Use Case: Online Boutique

Although there are numerous examples of components, services, and functionalities that can undergo testing through Chaos Engineering, our emphasis is on AIOps solutions. In the use case to be outlined in this subsection, the objective is to address two fundamental questions:

- How does AIOps handle a running chaos experiment?

- Does AIOps possess the capability to detect a running Chaos Experiment via observability?

Let us describe a concrete use case. The starting point is the Online Boutique store, which reflects our target system. Online Boutique is an open-source, cloud-native microservices demo application. Figure 8-2 illustrates the composition of the Online Boutique's architecture and its various microservices.

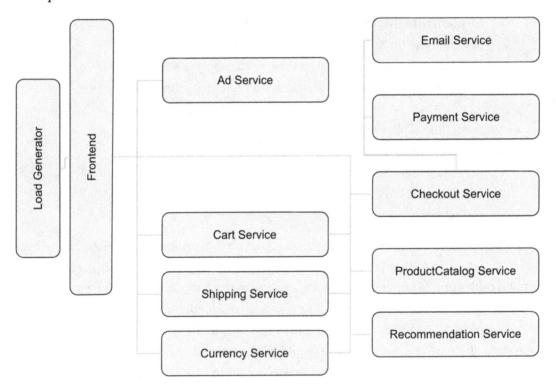

Figure 8-2. *Online Boutique Store Microservices Architecture*

The architecture encompasses all essential services needed for a standard e-commerce scenario, including frontend, payment service, checkout service, and more. This demo application can be easily pulled up locally with a Kubernetes cluster (e.g., by deploying a minikube instance). Once deployed, the Kubernetes cluster runs the 10 microservices, creating a simulation of a web-based e-commerce application.

Once all microservices are in the status *Ready* and *Running*, an observability tool is required to visualize the microservice architecture. Depending on the tool in question, there are various ways to onboard the relevant data into the observability platform. A common standard used today is OpenTelemetry, which can be used to collect and forward all Kubernetes telemetry (metrics, traces, logs). Nevertheless, there are also other technologies available on the market.

Common observability tools (e.g., Dynatrace, Splunk) have an **Application Performance Monitoring** (**APM**) feature, which allows them to map the target system component architecture to a user-friendly UI, which often features different levels of granularity (e.g., drill-downs). On top of that, such tools in many cases provide out-of-the-box AI capabilities, such as anomaly detection and root cause analysis, which can be used to detect and predict numerous types of failures by analyzing various historical patterns.

Before we start defining the actual steps for initiating a chaos experiment, it is necessary to identify the component or service that we want to target. Looking back at the microservice diagram, it is clear that a user cannot complete a purchase without triggering the "cartservice." Therefore, an interesting chaos experiment could entail the disruption of such service. If the "cartservice" were to fail, customers would not be able to complete their checkout, which would lead to a frustrating experience. To do so, one could, for instance, increase the latency for "cartservice" in order to test the resiliency of the system (in this case, by system, we refer to the AIOps solution, which we consider already built into our observability platform, which we are already using to gain insights into the application).

Listing 8-1. Example LitmusChaos YAML pod-network-latency

```
apiVersion: litmuschaos.io/v1alpha1
kind: ChaosExperiment
metadata:
  name: pod-network-latency
```

```
spec:
  definition:
    scope: Namespaced
    permissions:
      # RBAC details...
    image: "litmuschaos/go-runner:latest"
    args:
      - -c
      - ./experiments -name pod-network-latency
    env:
      - name: TOTAL_CHAOS_DURATION
        value: '60'
      - name: NETWORK_LATENCY
        value: '2000'
      - name: NETWORK_INTERFACE
        value: 'eth0'
    labels:
      name: pod-network-latency
```

Before running an experiment, it is often a good practice to create a detector. Hence, a "cartservice latency" detector has been implemented with the aim of monitoring the latency trends associated with the "cartservice." Depending on the observability tool in question, there are different types of anomaly detection algorithms from which one can choose. For this use case, a condition has been set up on the detector, which triggers an alert if the latency value is 3 standard deviations higher than the mean of the preceding hour. This kind of approach is useful for detecting unexpected increases (or spikes) in latency.

Listing 8-2. Example Dynatrace anomaly detection rule JSON

```
{
  "responseTimeDegradation": {
    "detectionMode": "DETECT_AUTOMATICALLY",
    "automaticDetection": {
      "loadThreshold": "ONE_REQUEST_PER_MINUTE",
      "responseTimeDegradationMilliseconds": 250,
      "responseTimeDegradationPercent": 90,
```

```
      "slowestResponseTimeDegradationMilliseconds": 500,
      "slowestResponseTimeDegradationPercent": 200
    }
  },
  "failureRateIncrease": {
    "detectionMode": "DETECT_USING_FIXED_THRESHOLDS",
    "thresholds": {
      "threshold": 10,
      "sensitivity": "LOW"
    }
  },
  "trafficDrop": {
    "enabled": true,
    "trafficDropPercent": 95
  }
}
```

During the execution of chaos experiments, our objective is frequently to replicate scenarios that closely resemble real-world conditions. The intention is to replicate a service disruption with a chaos experiment, mimicking its occurrence in a production environment, despite the chaos experiment being executed in a test environment. Simply deploying the Online Boutique is not sufficient; the crucial element missing is the presence of various users flooding the website, mirroring the dynamics of a production environment. In order to achieve this type of simulation, we will be making use of Locust, an open-source load testing tool, which can be used to continuously simulate various user actions on the Online Boutique. Locust can also be easily deployed via a local Kubernetes cluster. As shown in Figure 8-3, in Locust we can define two values:

- **Number of Users to Simulate**: As the name itself suggests, here we can define the number of concurrent users that Locust should simulate on the Online Boutique store.

- **Spawn Rate**: Refers to the pace at which users are initially generated before the specified number of concurrent users is reached.

Type	Name	# Requests	# Fails	Median (ms)	90%ile (ms)	Average (ms)	Min (ms)	Max (ms)	Average size (bytes)	Current RPS	Current Failures/s
GET	/	1613	0	370	2600	782	12	5168	10601	0.6	0
GET	/cart	4214	0	310	750	379	12	5025	14048	3.2	0
POST	/cart	4245	0	640	1200	680	18	5511	17453	3.6	0
POST	/cart/checkout	1355	0	1100	5800	6673	79	823518	7515	0.9	0
GET	/product/0PUK6V6EV0	2046	0	310	740	377	13	4394	8431	2.1	0
GET	/product/1YMWWN1N4O	1968	0	310	750	396	13	4991	8507	1.4	0
GET	/product/2ZYFJ3GM2N	2005	0	290	750	386	13	4725	8489	1.9	0
GET	/product/66VCHSJNUP	2029	0	300	740	375	13	5149	8492	2.5	0
GET	/product/9E92ZMYYFZ	2000	0	310	750	387	13	5041	8502	1.5	0
GET	/product/9SIQT8TOJO	2061	0	330	750	398	12	4324	8472	1.4	0
GET	/product/L9ECAV7KIM	2012	0	290	740	378	13	5042	8491	1.8	0
GET	/product/LS4PSXUNUM	2081	0	320	740	392	13	4203	8524	1.6	0
GET	/product/OLJCESPCTZ	2054	0	310	750	389	13	4127	8461	1.8	0
POST	/setCurrency	2820	0	330	760	404	15	4419	10606	2.3	0
	Aggregated	32523	0	380	930	707	12	823518	10625	26.6	0

Figure 8-3. *Example Locust UI Start Load Testing*

As we can see from the figure, Locust previews the set of simulated user actions in the form of HTTP requests. For each of these requests, it keeps track of a set of metrics, such as the number of requests of the same type that were executed ("# requests"), out of which it indicates the amount that failed ("# fails"), as well as some additional performance metrics: median, average, min, max, etc. This table provides a real-time overview of simulated user actions on the Online Boutique store. On the top right side of the figure, an interesting value is indicated, which is the failure percentage (35%). This figure indicates the failure rate of the executed requests on the site, serving as a valuable indicator of the website's overall health.

Now that we have a simulated environment and simulated users, we can move on to the generation of the actual chaos experiment. For that purpose, one could make use of LitmusChaos, an open-source, end-to-end Chaos Engineering framework intended for cloud-native applications.

Figure 8-4 illustrates the various steps required to create an experiment in Litmus Chaos:

1. **Schedule Workflow:** Once logged into the Litmus Chaos portal, in our test environment, which is `http://litmus-chaos.localdomain:8000/login`, select the **Litmus Workflow** tab to initiate the configuration of the chaos experiment. Select **Schedule a workflow.**

2. **Choose Agent:** Select the Kubernetes cluster used to deploy Online Boutique (**Self-Agent**).

3. **Workflow Settings:** There are various options for creating a workflow. It can be created from a template, cloned from an existing workflow, or imported using a YAML. If you are creating a workflow for the first time, select **Create a new workflow using the experiments from ChaosHubs**

4. **Tune Workflow:** In this phase, we define the type of chaos experiment that we want to run. We start by selecting **Add new experiment**, after which we obtain a list of different types of chaos experiments that we can inject into our target system, for example, pod deletion, container kill, CPU hog, network loss, etc. Since the goal is to inject latency issues into "cartservice," we will select **generic/pod-network-latency** from the list. It is possible to run multiple chaos experiments sequentially; however, for this use, we will only focus on one. Proceed by selecting the edit icon next to the experiment we just added (**pod-network-latency**) to further configure the experiment. Under **General**, we can leave the naming of the experiment as is. The more relevant section is **Target Application**, where under **applabel**, we will define **app=cartservice**, as the goal is to disrupt **cartservice** by introducing an increased latency. Following that, within the **Tune Experiment** section, we have the option to specify the duration of the chaos experiment, the precise network latency for injection, and the amount of jitter. Once the experiment has been configured, select **Finish**. Before further proceeding with the configuration of the workflow, it is highly suggested to edit **Advanced Options** and enable **Cleanup Chaos Workflow Pods**, which guarantees the cleanup and restoration of the environment once the chaos experiment has been completed.

5. **Reliability Score:** Provides the option to add a reliability score between 1 and 10 for each selected chaos experiment. Since we are running only one, this part can be skipped.

6. **Schedule:** One can schedule the configured experiment to run
   at a specific time of day (**Recurring Schedule**) or simply run the
   chaos experiment right away (**Schedule now**).

7. **Verify and Commit:** Summary of the chaos experiment and
   workflow configuration. Select **Finish** to initiate the workflow.

Figure 8-4. *LitmusChaos: Steps for Creating a Chaos Experiment*

Once the steps are completed, LitmusChaos will initialize and run the chaos
experiment.

Coming back to Locust, let us focus on Figure 8-5, where we can notice a significant
increase in failure rate. If we look at *"Failures"* on the top right, we will notice it is
already close to 60%. Additionally, notice that the highest number of failures is related
to requests that call the cart service (underlined in red). This is a good indicator that the
chaos experiment was successfully initiated and that it is already having an impact on
the target system.

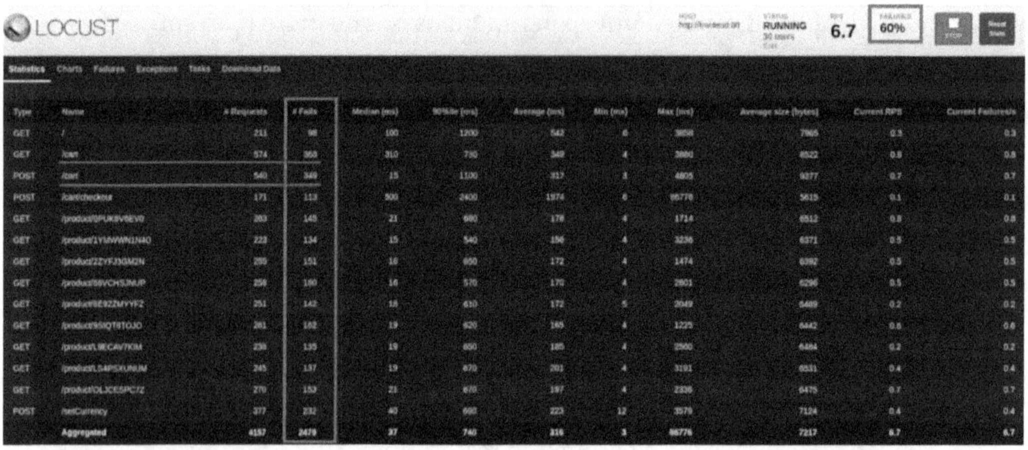

Figure 8-5. *Example Locust UI Before the Initiation of Chaos Experiment*

By going back to the observability tool and inspecting the latency detector for
cartservice, it is evident that an alert has been triggered. Notice in Figure 8-6 that
between approximately 14:20 and 15:10, the latency values have constantly oscillated
between 40 and 80 milliseconds. From approximately 15:10, the latency starts to increase

up to 120 milliseconds, which is significantly above the baseline. Additionally, notice that the alert was fired at 15:10, just a couple of minutes before the latency started to notably increase. What we observe here is the power of AIOps, capable of proactively identifying failures before they occur.

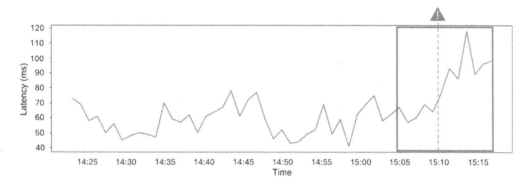

Figure 8-6. *Example Cartservice Latency Detector*

By enabling AIOps in a similar fashion, operation teams are able to significantly reduce the **mean time to detect** (**MTTD**). Most observability frameworks provide integrations with other communication platforms (e.g., email, Teams) or ITSM tools (e.g., ServiceNow). In this manner, an on-call **site reliability engineer** (**SRE**) could be immediately notified about the escalating latency issue, providing them with additional time to address the failure before its impact intensifies (or, in the realm of this use case, before it reaches the end-user). Furthermore, it is possible to take an additional step toward automation. Instead of simply alerting upon anomaly detection and only then applying a fix, one could implement a runbook (e.g., Rundeck, Ansible, RPA), featured by numerous observability frameworks, which executes an automated remediation script. This script has the potential to address the latency issue, such as restarting the node on the cart service. This is an example of a shift from reactive to predictive.

In addition to the latency detector, it is possible to construct a chaos experiment detector. Such a detector has the capability to utilize historical data and patterns, enabling it to identify periods when a chaos experiment is actively being conducted on the target system. In this particular use case, such a detector has been built, successfully identifying every instance in which LitmusChaos initiates a chaos experiment—essentially functioning as a chaos counter.

Initially, we have posed two questions, which we can now answer and elaborate on:

1. How does AIOps handle a running chaos experiment?

    Answer: As proven by the use case, if configured correctly and in the case of sufficient historical data, AIOps is capable of proactively detecting incoming disruptions on the target system. This allows operational teams the flexibility to address the issue before the impact escalates further or to implement automated remediation procedures through runbooks.

2. Does AIOps possess the capability to detect a running Chaos Experiment via observability?

    Answer: As proven by the use case, if the observability tool is collecting relevant and sufficient data on the target system, it is possible to build a chaos detector, which is able to identify each occurrence of a chaos experiment running on the target system. This is precisely how AIOps can be applied to leverage and fine-tune chaos experiments.

As we learned, AIOps plays a key role in refining chaos experiments by providing a sophisticated layer of intelligence to IT operations powered by AI/ML (artificial intelligence and machine learning). By harnessing AIOps to refine chaos experiments, organizations can significantly enhance their IT resilience through intelligent automation and predictive insights. This proactive stance allows for more precise and effective chaos experiments by targeting specific system behaviors and predicting their outcomes. With AIOps, organizations can dynamically adjust their chaos engineering strategies, ensuring that they not only react to known failure modes but also anticipate and mitigate unknown issues. By continuously learning from historical incidents and experiments, we can fine-tune the machine learning models used for AIOPS and also discover new unknown patterns.

Automation and Orchestration Techniques

In modern site reliability engineering (SRE), automation and orchestration are not just helpful add-ons; they are fundamental building blocks for achieving consistency, reliability, and scale. Automation means handling repetitive, predictable tasks with minimal human intervention, while orchestration connects these automated tasks into reliable end-to-end workflows that ensure systems remain resilient, even under stress.

One of the biggest benefits of automation is the reduction of toil, those repetitive, manual activities that drain valuable engineering time. By automating tasks like routine deployments, configuration updates, incident remediation, and system health checks, teams free themselves to focus on higher-value work, such as improving resilience patterns or experimenting with new failure scenarios. For example, automating container scaling through Kubernetes' Horizontal Pod Autoscaler ensures that applications can respond to sudden traffic spikes without an engineer needing to step in manually.

Orchestration builds on this by stitching automated steps together. This is especially powerful for CI/CD pipelines, incident response playbooks, or chaos experiments. A well-orchestrated pipeline might automatically deploy code, run tests, inject faults into a staging environment, monitor the results, and even roll back if critical health checks fail, all without human touchpoints. This tight integration of automation and orchestration builds confidence and creates faster feedback loops, which are essential in dynamic production environments.

To put these principles into practice, many teams use Infrastructure as Code (IaC) tools like Terraform or Pulumi to provision and manage infrastructure consistently, or configuration management tools like Ansible or Chef to keep environments in the desired state. Runbooks are increasingly codified and stored in source control, making incident responses more predictable and auditable. Chaos engineering tools such as Steadybit, LitmusChaos, or Chaos Mesh can be integrated into pipelines to ensure resilience is tested automatically as part of everyday workflows. CI/CD orchestrators like GitLab CI, Argo Workflows, and GitHub Actions help teams manage this complexity at scale.

A good policy to adopt is to regularly review operational tasks and identify candidates for automation and orchestration. Teams should revisit these opportunities every quarter to ensure they're not missing hidden toil. As organizations mature, measuring the impact of automation and orchestration becomes critical; key metrics include the percentage of toil reduced, improvements in deployment frequency, and reductions in mean time to resolve (MTTR) when incidents do occur.

In short, automation and orchestration are force multipliers. They reduce human error, increase speed and consistency, and create space for engineers to tackle more strategic challenges, turning the goal of resilient, reliable systems into an everyday reality.

Evaluating Trade-offs and Integrations

Now we want to provide strategies for organizations to maximize the adoption and utilization of SRE tools post-implementation. We discuss techniques for training, process integration, and fostering a culture of experimentation and resilience. We want to empathize with the strong need for training. Organizations have their employees, and to have an impact in the short term, we need upskilling. Each organization has its unique needs. First, we understand the organizational needs. For that, we begin by emphasizing the importance of understanding the unique needs and goals of the organization before rolling out our SRE tools. It is necessary to align tool adoption strategies with broader organizational objectives. We do this with prioritizations, and we want to tailor the approach for different business units. Table 8-1 presents the different persona involved in SRE adoption, outlining their primary concers and what they require from SRE tooling.

Table 8-1. *Example Persona involved in SRE adoption, primary concerns and SRE Tooling Needed*

Persona	Primary Concerns	What They Need from SRE Tooling
DevOps engineer	Software development, automation, toil reduction, seamless integration with pipelines, feature flag	Detailed runbooks, self-service automation scripts, good APIs, software architecture features to increase reliability, and feature deployment
Cross-functional site reliability engineer (SRE)	Reliability improvement, alerting, resilience validation, automation, toil reduction, preventive maintenance, observability instrumentation	Advanced dashboards, SLO error budget reports, chaos experiment templates, advanced automation orchestration, production readiness assessment, and dynamic tracing
Product owner	Feature delivery speed, trade-offs between reliability and velocity	Visibility into SLO compliance, clear status reporting, incident trends, incident risk, and business impact
Team lead/ manager	Cross-team adoption, ROI justification, upskilling gaps	Training materials, usage analytics, and clear value metrics
Business stakeholder	Customer trust, financial impact of outages, brand reputation	High-level summaries of uptime, incident cost trends, business risk reports, incident risk, and business impact

Often, we adjust the training program for the segment of our company slightly. Developing a comprehensive training program is significant to ensure that all relevant stakeholders, from developers to operations teams, are proficient in using SRE tools effectively. We want to focus on hands-on training and continuous learning opportunities in what we call "office hours," where practitioners can drop in with specific questions.

The moment we integrate SRE with existing processes, we see a strong push in the adoption. The seamless integration of our SRE tool into existing processes, such as CI/CD pipelines, monitoring and observability systems, and incident response protocols, helps practitioners to reduce the effort and increase the stability. It helps to provide practical guidance on how to facilitate this integration to minimize disruption. The interaction points, especially when multiple teams are involved, want to be laid out and thoughtfully designed.

We address the cultural shift required to foster a mindset of experimentation and resilience within the organization. It supports discussing strategies for promoting a culture where failure is embraced as an opportunity for learning and improvement. I have been in calls where the Head of Technology shared his moment of failure, even when it was years back. He was even talking about his feelings and embarrassment and switched at the end to his lesson. The learnings he applied after his day of failure.

This is a great example of leadership buy-in and support for a cultural shift. We cannot stress enough the importance of obtaining buy-in and support from organizational leadership to drive the adoption of a chaos engineering mindset, tools, and practice. Effectively communicating from top down and a shared understanding of the value proposition of chaos engineering will accelerate the adoption by gaining leadership endorsement.

Another element of a smooth adoption process is cross-functional collaboration. Fostering collaboration and communication between cross-functional teams involved in the adoption of SRE tools is significant. One team alone will always learn much slower, and the motivation will decrease compared to when you have multiple teams helping each other and driving in the same direction. We have multiple techniques to break down silos and promote collaboration across departments:

- **Cross-Functional Workshops:** Organize workshops where representatives from different departments come together to understand the concept of SRE.

- **Shared Goals and Metrics:** Establish shared goals and metrics that align with the organization's overarching objectives. When teams have a common purpose and understand how SRE contributes to achieving those goals, they are more likely to collaborate effectively.

- **Cross-Functional Teams:** Form cross-functional teams specifically dedicated to SRE initiatives.

- **Transparent Communication Channels:** Create transparent communication channels to facilitate information sharing and collaboration across silos.

- **Training and Skill Development:** Provide training and skill development opportunities related to SRE for employees across different departments. This ensures that everyone has the necessary knowledge and expertise to actively participate in SRE activities.

- **Incentivize Collaboration:** Implement incentives and recognition programs that reward collaboration and cross-functional teamwork.

- **Executive Support and Sponsorship:** Secure executive support and sponsorship for SRE initiatives. When senior leaders endorse and actively promote SRE and the collaboration across organizational silos, it sends a powerful message throughout the organization.

We want to establish a good feedback loop, including sharing the feedback from the tool with the vendor. Normally, it's best to check in after the first steps with the tool and get a clear understanding of where the hurdles are. We document this transparently and share it not only with the team but also with our partners. When we address the feedback with a change in the tool or the process, most companies forget to share this back to the team that raised the issues.

Recognition and incentives for teams doing SRE are two relevant steps to increase the adoption of our new tool. Celebrating successes, acknowledging contributions, and providing incentives for participation, especially when it comes from leadership.

If we do all of this, we will see increased community engagement and knowledge sharing between the teams. They will start to exchange knowledge, best practices, and lessons learned. We will see that more mature questions are being discussed in our forums and that employees strive to attend conferences and community events.

Table 8-2 presents a list with relevant tools for each capability. It is not a comprehensive list; it should just give an indication about options to consider.

Table 8-2. *Example SRE Tools and Capabilities*

Capability Area	Example Tools
AIOps	Dynatrace, DataDog, Moogsoft, Splunk ITSI, PagerDuty AI Ops
Automated on-call support	PagerDuty, Opsgenie, VictorOps, xMatters, Squadcast
SLOs and Error Budget	Nobl9, DataDog, Lightstep, Blameless, Google Cloud Operations Suite (formerly Stackdriver)
Observability	DataDog, Elastic Observability, Splunk, Dynatrace, Grafana, Prometheus
Feature flags	LaunchDarkly, Split.io, Harness, Flagsmith, FeatureFlow
Pre-mortem/What-If scenario analysis	Miro, Mural, Confluence (for documentation), ChaosIQ, AWS Fault Injection Simulator (FIS)
Critical user journey identification	WalkMe, Heap, Mixpanel, Pendo, FullStory
Synthetic monitoring	Pingdom, Catchpoint, Uptrends, Selenium, Dynatrace synthetic, Splunk synthetic
Reliable experiments	Optimizely, Harness, LaunchDarkly, Argo Rollouts (Kubernetes), Split.io
Chaos testing and engineering	Gremlin, Litmus Chaos, AWS Fault Injection Simulator, Chaos Mesh, Azure Chaos Studio, Steadybit, Harness
Toil management	Jira (for toil tracking), Rundeck, ServiceNow Automation, Jira Service Management, Ansible, Puppet, Terraform
Configuration management	Ansible, Chef, Puppet, Terraform
Capacity management	AWS Trusted Advisor, DataDog, VMware vRealize Operations, Turbonomic, Azure Monitor
Blameless postmortem	Confluence, Blameless, Miro, Google Docs
Automated risk assessment in change management	Digital.AI (Numerify), ServiceNow Change Management, GitOps with ArgoCD, Harness Continuous Verification, Jira Service Management
Build-run teams	N/A (Team Organization Structure)
Psychological safety	InsightScan, Officevibe

Diving Deep into Data Collection, Analysis, and Visualization for Site Reliability Engineering

At the heart of SRE lies our data-driven approach, which relies on meticulous data collection to inform decision-making and experimentation. This method entails systematically gathering, analyzing, and leveraging data to gain insights into system behavior, identify weaknesses, and drive continuous improvement. Let's look deeper into the framework of a data-driven approach in chaos engineering.

Data collection serves as the foundational pillar of Chaos Engineering. It involves gathering a diverse range of data from multiple sources, including user interactions, network traffic, and system performance metrics. This comprehensive data collection provides the necessary insights into the intricacies of the system under study. To do this effectively, many teams use tools like **Fluent Bit** for streaming logs, **Jaeger** for distributed traces, and **Prometheus** for time-series metrics. This ensures that even edge cases and failure scenarios, which may not be captured in traditional monitoring, are recorded in formats like structured logs (JSON), traces, and metrics, giving teams the granular, low-level visibility needed to understand real system behavior under stress.

Informed by the data collected, we designed the experiment with the aim of identifying key system components, failure modes, and performance metrics to be targeted. By leveraging insights gleaned from data analysis, experimental design ensures that chaos experiments focus on the most critical aspects of the system, leading to more meaningful insights and improvements. We want to have the experts for the system by hand to understand system architecture and interdependencies, allowing them to design experiments that target the most critical failure points and uncover hidden vulnerabilities. They can anticipate and simulate complex failure scenarios that may not be obvious to less experienced practitioners.

Our comprehensive data collection enables the quantification of the impact of chaos experiments on system resilience, performance, and overall health. This step involves analyzing the data to assess the effectiveness of chaos engineering practices and validate their impact on the system. When we do this regularly, we understand our impact, and we can inform practitioners and build an enterprise-wide data collection. This involves the data we collect during experiments in correlation with business-level metrics and KPIs, enabling them to quantify the impact of chaos experiments in a way that resonates with our stakeholders. We want to understand the importance of establishing baselines and measuring the long-term effects of chaos experiments on system resilience. This is the difference between having chaos teams and a chaos enterprise. We only see real improvements for our company if we are able to level up from teams to our whole enterprise.

For that, data analysis plays a pivotal role in deriving actionable insights from the collected data. Advanced analytical techniques, such as anomaly detection and preventive/predictive analysis, help uncover underlying patterns and relationships that inform remediation strategies and drive continuous improvement. To do this well, we want our experts to be skilled in applying advanced methods, including machine learning and statistical modeling, to detect subtle patterns that may not be immediately apparent. This allows them to identify complex, multivariate dependencies that can contribute to system failures. When we analyze data across multiple systems, we gain a clearer understanding of interdependencies and can improve overall reliability. It's important to remember, however, that not all observability tools support cross-service correlation out of the box; achieving this often requires custom logic, tailored dashboards, or additional integration to stitch together traces, logs, and metrics from disparate services.

A key aspect for communication but also for analytics is that we visualize our data. Effective data visualization is essential for interpreting chaos experiment results and communicating insights to stakeholders. Multiple tools and techniques can be used. Visualization tools, such as heatmaps and interactive dashboards, provide a clear and intuitive representation of system performance metrics and failure data, facilitating data-driven decision-making. We want in our training to teach this skill; we have seen communication in the right way accelerates the adoption and our practice quickly. We want our practitioners to be ready to design intuitive and informative visualizations that effectively communicate the insights derived from chaos experiments. What we learned is that tailoring the visualization is key. Each stakeholder has specific needs and perspectives, from developers to business leaders.

Table 8-3. *Example Visualization Dashboard Heatmap and Interactive with Cross Cursor*

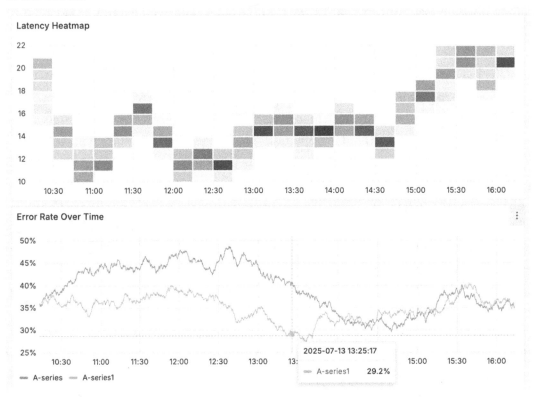

The data collection and the visualization come into play again during the post-experiment analysis. It's critical for identifying systemic weaknesses and informing future experiment design. By analyzing the data collected and deriving actionable insights, our teams prioritize improvement efforts and drive continuous refinement of the chaos engineering process. We extract the maximum value from post-experiment analysis by combining the data collected during the experiment with historical data and contextual information. With visualization and historical data from our collection, we identify subtle, long-term trends and patterns that may not be immediately apparent.

This gets more effective when we collaborate across different teams within the organization, and all are using the same data-driven approach. By fostering a shared understanding of system behavior and failure modes and democratizing data-driven insights, teams can drive continuous improvement and enhance system resilience and performance. When using the same visuals and using a similar data-driven approach, we are fostering a culture of shared responsibility and continuous learning across different teams. This helps us to break down silos and align diverse perspectives from our practitioners across different business units towards a common goal of improving system resilience.

Chaos engineering is an iterative process that adapts to evolving system complexity. By continuously refining data collection, analysis, and experimentation strategies, teams can drive ongoing improvements to the system, enhancing its overall resilience and performance.

Summary
Scaling with Purposeful Tooling

Throughout this chapter, we've unpacked the critical role tools play in scaling SRE practices effectively and sustainably. We began by examining how to assess enterprise tooling needs, not just from a technical fit perspective, but through a strategic lens that aligns with organizational priorities and maturity.

We then explored advanced tooling categories that underpin modern reliability engineering: chaos experimentation frameworks that expose system weaknesses before they escalate, observability platforms that surface insights from deep within distributed systems, and AIOps solutions that bring machine intelligence to noise reduction and anomaly detection.

On the execution front, we studied how automation and orchestration unlock operational scale, reduce toil, and promote consistency across environments, whether managing incidents, deploying code, or provisioning infrastructure.

Finally, we emphasized the importance of making deliberate trade-offs. No tool exists in isolation, and understanding integration complexity, team readiness, and long-term scalability is as vital as the tool's feature set itself.

By the end of this chapter, we've not only learned how to select tools, but you've also learned how to strategically scale reliability through thoughtful, contextual tooling choices that elevate both people and platforms. As we move into the post-rollout phase, remember to apply best practices such as **monitoring feature adoption through a license usage dashboard** and **reviewing tool usage patterns regularly** to ensure that teams are actually using the capabilities you've invested in, and adjusting training or configuration as needed to maximize impact.

## CHAPTER 9

Influence of AI and Generative AI in SRE Adoption

In this chapter, we examine how artificial intelligence, particularly predictive, causal, and generative models, is influencing the adoption and evolution of Site Reliability Engineering (SRE) practices. As systems become increasingly distributed, fast-moving, and unpredictable, traditional automation and monitoring approaches are often insufficient for maintaining reliability at scale. AI is emerging not merely as an enhancement to SRE workflows but as a foundational capability: augmenting human decision-making, enabling more adaptive automation, and generating operational knowledge directly from telemetry and historical data.

We begin by exploring the broader role of AI in modern SRE, focusing on how it supports core principles such as embracing risk, reducing toil, and improving decision-making. We then dive into the growing use of predictive and causal models to detect emerging issues and trigger automated remediation workflows—an evolution toward more proactive and context-aware operations. Finally, we look at the rising influence of Generative AI, particularly in automating the creation of documentation and runbooks, and how this capability is helping teams keep pace with system complexity while capturing institutional knowledge more effectively. We close with a forward-looking perspective on how these trends may reshape the way reliability engineering is practiced in the years ahead.

© Florian Hoeppner, Francesco Sbaraglia 2025
F. Hoeppner and F. Sbaraglia, *Mastering Site Reliability Engineering in Enterprise*,
https://doi.org/10.1007/979-8-8688-1448-8_9

This chapter will cover the following main topics:

- **Role of AI in modern SRE**

- **Predictive analysis and automated remediation**

- **Generative AI for documentation and runbooks**

- **Agentic AI for SRE**

Role of AI in Modern SRE

With today's ever-growing complexity of modern, dynamic, and hybrid systems, it's becoming increasingly challenging for SREs to ensure reliability at scale. Frequent infrastructure changes, rapid deployments, multi-cloud environments, and loosely coupled services all contribute to a fragile operating landscape where pinpointing and resolving issues quickly is more complicated than ever. On top of that, teams are often forced to juggle disjointed tools and processes just to stay on top of what's happening in their systems.

Automation remains a foundational pillar of SRE practice, crucial for eliminating toil, enforcing consistency, and enabling scalable operations. However, in increasingly unpredictable and fast-moving environments, automation alone can reach its limits. Static rules and predefined responses are often insufficient when signals are noisy and failure modes are unfamiliar. This is precisely why AI is evolving from a helpful tool into an essential component of the modern SRE stack. As systems grow in scale and dynamism, AI augments human judgment and extends the reach of automation, enabling the surfacing of hidden patterns, the prioritization of critical signals over noise, and continuous adaptation to evolving system behaviors—capabilities that static tooling inherently struggles to provide.

In the contemporary landscape, where systems are increasingly characterized by their scale, dynamism, and unpredictability, traditional automation and monitoring approaches often prove insufficient for maintaining reliability. This is precisely why AI has emerged as an essential enhancement to the modern SRE stack. AI extends the capabilities of automation and augments human judgment, offering the capacity to uncover hidden patterns, prioritize critical signals over noise, and continually adapt to evolving system behaviors—functionalities that static tooling struggles to provide.

In contemporary implementations, AI engines are deeply integrated into modern SRE platforms, commonly seen within observability tools such as Datadog, Dynatrace, or Splunk AIOps, and extending to ITSM systems like ServiceNow and automated on-call support tools like PagerDuty AI Ops. These AIOps solutions apply machine learning and data science to IT operations, leveraging big data, analytics, and AI techniques to automate the identification and resolution of common IT issues. They achieve this by analyzing vast volumes of historical operational data, interpreting golden signals (metrics, logs, traces), and identifying patterns that human operators might miss, thereby enabling more proactive and intelligent decision-making. This robust integration empowers SREs to detect risks earlier, automate responses intelligently, and accelerate incident response in systems too complex for manual management, shifting from a reactive to a proactive stance by leveraging AI-driven insights that continually evolve with system behavior.

In systems that are increasingly complex and challenging to manage manually, AI provides SREs with the crucial capacity to detect risks earlier, automate intelligently, and accelerate incident response. Moving beyond static rules or predefined dashboards, teams now leverage AI-driven insights that evolve with system behavior, fostering a proactive rather than reactive stance. This evolution is significantly driven by AIOps, which applies machine learning and data science to IT operations, automating the identification and resolution of common IT issues by analyzing large volumes of historical operational data to predict and prevent problems. These AI capabilities are commonly integrated into various platforms within the modern SRE stack: for instance, observability tools like Datadog, Dynatrace, and Splunk frequently incorporate AI engines for anomaly detection and root cause analysis. Similarly, ServiceNow (an ITSM system) can receive notifications and trigger automated remediation workflows from AI engines, and PagerDuty AI Ops is listed among automated on-call support tools that leverage AI to classify tickets, prioritize alerts, and summarize incidents. This integration empowers SREs to manage highly dynamic and unpredictable environments by intelligently identifying patterns and taking action before user impact. No specific version information for these platforms is available in the provided sources.

Let's take a look at how AI is being applied across the core principles of modern SRE practice.

1. **Embrace Risk**

    SRE operates from the fundamental premise that 100% reliability is neither realistic nor always desirable. Every system inherently involves a trade-off between feature velocity and operational stability, and effectively managing this dynamic necessitates a deep understanding and embrace of calculated risk. This goes beyond merely setting Service Level Objectives (SLOs); it requires continuously evaluating the actual risk exposure of a system at any given moment. AI plays an increasingly vital role by enabling teams to quantify this risk with greater precision. For instance, machine learning models can forecast error budget consumption based on current trends, flag when a release might push a service beyond acceptable thresholds, or predict the likelihood of cascading failures given recent system behavior. These actionable insights, typically derived from specialized tooling and AIOps solutions like Nobl9 or those supporting OpenSLO concepts, empower SREs to make informed, data-backed decisions about when to accelerate innovation and when to prioritize stability.

2. **SLOs Are the Foundation**

SLOs guide critical decisions about reliability, risk, and engineering priorities. But in complex, distributed systems with constantly shifting dependencies, defining meaningful SLOs is not always straightforward. AI helps teams navigate this complexity by analyzing historical behavior, incident patterns, and user impact across varied topologies. Predictive models can simulate how different SLO targets might play out under real workloads. At the same time, causal inference techniques help identify which components actually influence user-facing reliability, cutting through noisy correlations. The result is a more informed, data-backed approach to setting SLOs that reflect both system behavior and user expectations, even in highly dynamic environments.

Predictive models, which often leverage time-series forecasting techniques (such as those employed in frameworks like Prophet or more advanced machine learning models like XGBoost, are instrumental in achieving this predictive capability. These models necessitate a substantial volume of high-quality historical telemetry data, including metrics, logs, and traces—collected over extended periods (e.g., weeks or months) to accurately identify patterns, learn system behavior under varying loads and conditions, and differentiate normal operating conditions from subtle anomalies that could lead to SLO breaches.

3. **Eliminate Toil**

   Toil—manual, repetitive, automatable work—drains time and
   focus from higher-value engineering efforts. Automation has
   always been the go-to strategy for reducing toil, but as systems
   grow in scale and complexity, not all toil is predictable or neatly
   scriptable. This is where AI steps in. It can classify tickets,
   prioritize alerts, summarize incidents, or detect recurring patterns
   in operational tasks that aren't obvious to humans. Over time, it
   learns from real workflows and adapts to changing conditions,
   handling edge cases that would require constant tuning in
   traditional automation. In short, AI helps eliminate not just the toil
   we know, but the toil we didn't even realize we had.

   For instance, AI-driven alert deduplication algorithms can
   identify and consolidate multiple, seemingly distinct alerts that,
   to a human operator, would trigger redundant escalations or
   investigations. By automatically recognizing underlying patterns
   and correlating these alerts, AI uncovers this "implicit toil" of
   unnecessary human intervention, thereby streamlining incident
   response and reducing alert fatigue for SRE teams.

4. **Measure Everything**

   Reliable systems depend on good data. Measurement informs
   everything from incident response to capacity planning to long-
   term reliability strategy. But as systems become more distributed,
   dynamic, and interdependent, simply collecting data isn't enough.
   AI helps make sense of it—connecting the dots across layers of
   infrastructure, services, and user impact. For example, AI can
   build context-aware system models by analyzing observability
   signals, dependency graphs, and historical incident data. These
   models often span both *vertical* (infrastructure-to-application)
   and *horizontal* (service-to-service) dimensions, revealing how
   failures propagate and where intervention has the most impact.
   Causal inference techniques go a step further, helping teams
   distinguish between surface-level symptoms and underlying

causes. With this kind of system understanding in place, downstream tasks like alert routing, automation, and remediation become far more effective, while also enabling faster, more accurate root cause analysis.

5. **Automate This Year's Job Away**

   SRE encourages a mindset of continuous automation, not just eliminating toil but evolving roles by handing off yesterday's manual tasks to systems. AI takes this further by enabling automation that's adaptive, context-aware, and capable of handling ambiguity. Instead of scripting every step, AI can learn from past actions and system behavior to decide how to respond in novel scenarios. For example, it can recommend or even trigger remediation actions during incidents, based on patterns it has seen across similar events. As systems change, AI-driven automation keeps pace—helping teams automate not just the routine but the nuanced. That said, boundaries are essential. In most cases, the automation itself is predefined—engineers determine what actions should be taken, while AI focuses on identifying *when*. A typical example is using AI to analyze observability signals and trigger a self-healing script when early signs of failure or performance degradation are detected. The logic—what to do and under what conditions—is still human-defined. AI's role is to recognize the right moment to act based on patterns in metrics, logs, or traces. This balance ensures automation remains deliberate while benefiting from AI's ability to operate in real time and at scale.

6. **Reduce Organizational Silos**

   SRE has always emphasized collaboration across teams—breaking down walls between development, operations, and product. AI can support this not just by providing insights but also by making those insights accessible and actionable across functions. When systems are complex and ownership is fragmented, AI can help build shared context by identifying cross-team dependencies, surfacing relevant historical incidents, or even translating technical patterns into a language that product or support teams can understand. For example, during an incident, an AI system might correlate a frontend performance drop with a backend configuration change, flag the affected services, and suggest which teams to notify. Instead of teams working in isolation, AI can help orchestrate a coordinated response, backed by shared data and system understanding.

7. **Use Data to Drive Decisions**

    Automation is only as effective as the decisions behind it, and decisions are only as good as the data they're based on. Once AI has helped make sense of complex signals and mapped relationships across systems (*Measure Everything*), it naturally extends into guiding action. SREs constantly face trade-offs: whether to accept a risk, delay a release, allocate effort toward automation, or raise the bar on an SLO. AI supports these decisions by turning raw telemetry into actionable insights— simulating outcomes, forecasting impact, and identifying where engineering effort will have the highest leverage. For example, instead of guessing which recurring incident type to prioritize, teams can use AI to weigh factors like frequency, resolution time, and user impact. Or, when planning a system redesign, predictive models can expose long-term reliability costs that wouldn't surface in short-term testing. In this way, AI shifts the role of data from something teams *look at* to something they can *act through*, supporting faster, more confident, and more strategic decisions.

    To summarize, AI is rapidly becoming part of the SRE toolbox— not as a replacement for human intuition and engineering, but as a force multiplier. From guiding decisions to powering smarter automation, its value grows with system complexity. Still, applying AI in production environments requires care. Models must be interpretable, data must be trustworthy, and the limits of automation must be respected. The most effective SRE teams will be those that treat AI as a collaborative layer: one that augments human expertise, scales decision-making, and brings clarity where complexity once dominated. As systems continue to evolve, so will the partnership between SREs and the intelligent systems that support them.

Predictive Analysis and Automated Remediation

As systems grow more distributed and dynamic, reactive incident handling is no longer enough. Modern SRE practice is shifting toward prediction, detecting issues early, and resolving them automatically before users are impacted. This evolution is made possible by combining *observability* with *AIOps*. Observability provides the relevant contextualized data, while AIOps brings the intelligence to interpret it, find patterns, and take action. The key enabler is context: topology-aware observability gives AI the ability to understand not just what is happening, but where and why. This allows us to move from alerting on symptoms to anticipating causes and, ultimately, triggering proactive remediation workflows.

Causal AI plays a central role in this shift. By learning the relationships between components, rather than just tracking correlations, systems can be modeled both vertically and horizontally. A vertical topology might trace how an application depends on containers, clusters, and underlying hosts. A horizontal topology follows service call chains and transactional dependencies across the stack. When these views are combined, we gain end-to-end insight into the system. This context is essential for meaningful predictive analysis: it's not enough to know that a metric spiked—we need to understand what that means in terms of impact, propagation, and downstream risk.

Figure 9-1 illustrates this idea of context-aware topologies. Causal AI leverages this structure to identify which components are likely sources of degradation and how failures could cascade. It helps SRE teams reason not just about what broke, but what could break next. For example, the system might detect that high I/O wait time on a specific node, caused by underlying disk contention, is likely to cause increased response times in a dependent authentication service within the next 10 minutes. Because these components are causally linked in both the vertical and horizontal topology, AI is able to connect infrastructure-level symptoms to user-facing service degradation before it fully manifests.

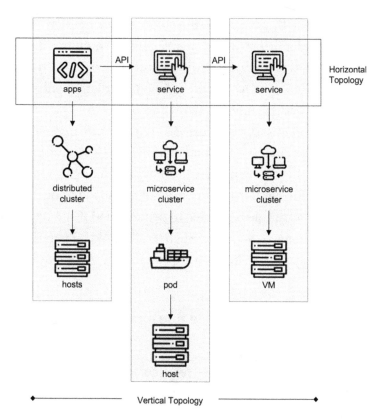

Figure 9-1. *Example of AI Context-Aware Topologies*

Once this level of insight is established, predictive models can identify early indicators of potential failures or performance regressions. A pattern of memory pressure might regularly precede pod evictions and user-facing latency. A queue growing in one service could be a known precursor to SLO burn rate in another. These insights are not deterministic—they're probabilistic, learned from data. But they provide enough lead time to act before an issue becomes critical.

What makes this especially powerful is the ability to link prediction directly to action. In modern observability platforms, predictive insights and anomaly detections can serve as entry points into AIOps workflows. These workflows follow defined rules—based on tags, namespaces, or service metadata—and initiate remediation only when conditions are met. For example, a workflow might be configured only to run a database connection pool reset script if the affected service is tagged as critical=true *and* belongs to the payments namespace. This ensures that remediation is only triggered in the correct operational context, avoiding unintended side effects and allowing teams to scope automation precisely.

What makes this especially powerful is the ability to link prediction directly to action. In modern observability platforms, predictive insights and anomaly detections can serve as entry points into AIOps workflows. These workflows follow defined rules—based on tags, namespaces, or service metadata—and initiate remediation only when conditions are met. For example, a workflow might be configured only to run a database connection pool reset script if the affected service is tagged as critical=true *and* belongs to the payments namespace. This ensures that remediation is only triggered in the correct operational context, avoiding unintended side effects and allowing teams to scope automation precisely. To further illustrate a basic AIOps remediation rule, consider a conceptual example where a rule is defined to restart a pod if latency exceeds two seconds within the 'payments' namespace. Such a rule could be structured in a format similar to the following JSON snippet, ensuring automated action based on specific conditions:

Listing 9-1. Example of an AI-generated rule name pod restart

```
{
  "ruleName": "High Latency Payments Service Remediation",
  "triggerConditions": {
    "metric": "latency",
    "threshold": "2s",
    "operator": ">",
    "serviceMetadata": {
      "namespace": "payments"
    }
  },
  "remediationAction": {
    "type": "restart_pod",
    "target": "affected_service"
  },
  "description": "Restarts pods in the 'payments' namespace if latency
exceeds 2 seconds."
}
```

This conceptual example is provided for illustrative purposes to demonstrate the potential structure of such a rule.

Upon detecting a problem, an AI engine—found in platforms like Moogsoft, BigPanda, Dynatrace, DataDog, Splunk ITSI, or PagerDuty AI Ops—can trigger a workflow to execute two parallel actions: first, notifying the SRE team via a ServiceNow ticket or an alert through systems such as PagerDuty, Opsgenie, VictorOps, xMatters, or Squadcast; and second, launching a self-healing script, which might involve autoscaling cloud resources or restarting a failing service. These automated workflows often leverage robust API support, including REST, CLI, and YAML integration, to ensure seamless and structured communication between disparate tools and systems.

Once the affected SLO stabilizes, the platform automatically updates the incident with resolution details. This is what we refer to as zero-touch operations—incidents that are detected, diagnosed, and resolved without human intervention, unless escalation is required. It reduces time-to-resolution, scales incident response, and frees up engineers to focus on higher-value work.

This automated process extends to IT Service Management (ITSM) tools like ServiceNow or Jira, where the AI engine not only creates an incident upon detection but can also dynamically update it with resolution details and automatically transition its status to 'Resolved' or 'Closed' once the self-healing actions prove successful and the SLO stabilizes. To illustrate this practical application, a conceptual example of a webhook payload that an AIOps system might send to an ITSM platform to update an incident with resolution details could resemble the following (note: this specific payload structure is illustrative and not directly from the provided sources):

Listing 9-2. Example of incident report payload JSON format

```
{
  "incident_id": "INC00012345",
  "status": "Resolved",
  "resolution_notes": "Automated remediation executed by AIOps agent.
  Identified database connection pool exhaustion on 'PaymentService'
  (triggered by increased traffic anomaly). Self-healing script
  successfully reset the connection pool. Service SLO ('Successful
  Payments') is now stable and within target thresholds. No human
  intervention required.",
  "resolved_by": "AIOps Automation Agent",
  "root_cause_summary": "Transient database connection pool exhaustion due
  to traffic surge.",
```

```
    "affected_service": "PaymentService",
    "resolution_timestamp": "2025-03-15T10:30:00Z",
    "slo_compliance_status": "Within Target"
}
```

Such a structured update ensures comprehensive, real-time documentation of automated resolutions, reducing manual overhead for SRE teams and providing clear audit trails for compliance and continuous learning, aligning with the broader goal of automating documentation and operational knowledge.

Figure 9-2 illustrates an example of a workflow in action. The observability platform continuously monitors the end-to-end target system. AI engines are responsible for detecting anomalies, building causal models, and identifying emerging risks based on complex patterns. In parallel, rule-based mechanisms monitor fixed thresholds, such as SLO breaches or hard KPI violations. When either type of trigger fires, a logical AIOps workflow is activated, which splits into two coordinated paths:

1. **Notification Path:** An incident is created in an ITSM system like ServiceNow, giving the SRE team full visibility into the issue. The incident includes contextual details such as affected services, probable root cause, and severity level. This ensures the team is kept in the loop from the beginning, even if no immediate human action is needed.

2. **Remediation Path:** At the same time, a pre-approved self-healing script is executed. This could be anything from autoscaling cloud resources to restarting a container to rolling back a faulty deployment. The remediation action is chosen based on the context and scope defined in the workflow rule.

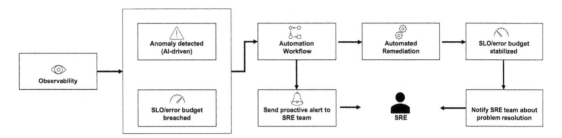

Figure 9-2. *Example of Zero-Touch AIOps Workflow in Action*

The workflow then enters a monitoring state, where it observes the system to confirm whether the issue has been resolved, typically by tracking whether the SLO has returned to a compliant state. If so, the workflow automatically updates the original incident with resolution details and closes the loop. If not, the incident remains open and may escalate to a human operator.

To put this into context, imagine a misconfigured SQL query deployed in production begins increasing response times and error rates for the authentication service. The AI engine detects the anomaly, identifies the historical link between similar query misconfigurations and degraded login performance, and fires the workflow. A ticket is created in ServiceNow with full diagnostic context, and in parallel, a self-healing script rolls back the misconfiguration and restarts the affected service. Once the system verifies that the login success rate has returned to normal and the error budget is no longer burning, the incident is closed automatically. The entire process—from detection to resolution—requires no human intervention. That's the essence of zero-touch AIOps: fast, intelligent, closed-loop automation that enhances reliability without increasing operational burden.

However, this zero-touch operation is augmented by intelligent guardrails and exception triggers to ensure reliability and trust in production environments. If the AI engine, through its predictive models and causal analysis, detects an anomaly but possesses a low confidence level in its proposed remediation, or if the automated self-healing script fails to stabilize the affected Service Level Objective (SLO), the incident is automatically escalated to a human operator. This human-in-the-loop mechanism ensures that while the system aims for maximum automation, human expertise and judgment are engaged when the AI is unsure or when the pre-approved automated actions are insufficient, maintaining deliberate and trustworthy automation.

Generative AI for Documentation and Runbooks

So far in this chapter, we've explored how predictive and causal AI are reshaping the core of modern SRE—from detecting anomalies before they escalate to enabling zero-touch remediation through intelligent automation. But while these AI models focus on real-time signal interpretation and decision-making, we shouldn't overlook the growing influence of Generative AI in shaping the knowledge layer that underpins operational excellence. Unlike traditional AI, which excels at classification and prediction, Generative AI is optimized for creation, turning raw inputs like logs, traces, tickets, and even natural language conversations into coherent, structured output.

In the context of SRE, Generative AI is being used in a variety of ways: drafting postmortems, summarizing incident threads, auto-generating changelogs, and even synthesizing architectural overviews.

GenAI is optimized for creation, transforming raw inputs such as logs, traces, tickets, and natural language conversations into coherent, structured output. Its particularly impactful applications lie in the creation, maintenance, and accessibility of documentation and runbooks, which have historically faced challenges with scale, consistency, and freshness. This technology automates tedious tasks, accelerates context creation, and embeds institutional memory directly into workflows.

For these generative capabilities, the process is typically grounded in a reliable data substrate, where observability plays a foundational role. Telemetry sources—including metrics, logs, traces, topology graphs, and event streams—provide both raw signals and the structural and temporal context necessary for meaningful documentation. When insights from predictive and causal AI models (which identify risks and root causes) are fused with this data, GenAI can compose structured, actionable narratives.

Specifically, GenAI is used to

- **Draft Postmortems**: Incident response data, encompassing alerts, logs, chat transcripts, and remediation actions, can be automatically distilled into comprehensive postmortems that include timelines, impact assessments, and contributing factors. These reviews are non-punitive and forward-looking, focusing on systemic improvement and learning from failures.

- **Summarize Incident Threads**: It helps condense complex incident discussions into concise summaries.

- **Auto-generate Changelogs**: Deployment metadata can be used to generate changelogs that highlight significant service updates or configuration changes, tailored for different audiences like engineering, QA, or compliance.

- **Synthesize Architectural Overviews**: Live service topology graphs can be used to generate architectural overviews that accurately reflect the current system composition, rather than relying on outdated diagrams.

- **Create Pre-mortem Reports**: Predictive AI models can analyze telemetry to identify emerging risks, and causal AI interprets the context to link potential issues to root causes. Generative AI then synthesizes this information into structured pre-mortem reports, providing high-confidence summaries of risks, expected behaviors, and recommended mitigation steps before an incident impacts users.

- **Generate Runbooks**: Generative AI can extract resolution patterns from past incidents, correlate them with telemetry, and formalize them into clear, repeatable, step-by-step procedures for operational tasks or incident response. These runbooks can become living operational assets, automatically generated, updated, and distributed as part of the platform. This also includes generating onboarding runbooks for new services or maintenance procedures.

- **Produce Periodic Operational Summaries**: Weekly or monthly reliability reports, which are often manually compiled, can be automatically generated by aggregating key metrics, Service Level Objective (SLO) trends, and incident patterns into a cohesive summary.

As described, these capabilities are increasingly integrated into modern observability platforms and other SRE/DevOps workflows. Several vendors are beginning to offer AI-assisted runbook generation or dynamic incident reporting directly tied to telemetry. This indicates a shift where operational knowledge is produced as a byproduct of system activity, rather than through after-the-fact manual documentation. Tools like Microsoft Copilot, Jasper, Notion AI, or custom LLM integrations in Datadog Notebooks are supporting these features. However, one particularly impactful application is in the creation, maintenance, and accessibility of documentation and runbooks—two foundational elements of reliability engineering that have historically struggled with scale, consistency, and freshness. As systems grow in complexity and incidents become more nuanced, the ability to keep operational knowledge accurate, discoverable, and actionable becomes just as important as responding to the incidents themselves. This is where Generative AI is beginning to make a tangible difference: not by replacing human expertise, but by capturing and amplifying it—automating the tedious, accelerating the creation of context, and embedding institutional memory into workflows where it's needed most.

To generate accurate and contextually relevant operational knowledge, GenAI must be grounded in a reliable data substrate. This reliable data substrate is formed by telemetry sources such as metrics, logs, traces, topology graphs, and event streams, which provide not only raw signals but also the structural and temporal context required for meaningful documentation. To ensure these observability signals are clean and ready for the Generative AI engine, data engineering pipelines can be employed. For instance, tools like Kafka can be used to stream real-time telemetry data, which is then processed and stored in systems such as Elasticsearch for efficient indexing and retrieval. This cleaned and structured data then serves as the input for LLMs within the Generative AI system. This robust foundation allows Generative AI systems to reconstruct the operational timeline of an event, infer contributing factors, and synthesize structured documentation automatically and at scale, transforming raw data into actionable operational knowledge.

This is where observability plays a foundational role. Telemetry sources such as metrics, logs, traces, topology graphs, and event streams provide not only raw signals but also the structural and temporal context required for meaningful documentation. Generative AI systems can use these signals to reconstruct the operational timeline of an event, infer contributing factors, and synthesize structured documentation— automatically and at scale.

Importantly, Generative AI is most effective when operating in tandem with other forms of AI. Predictive models surface forward-looking risks before symptoms manifest, while causal inference identifies the most likely failure paths based on dependency structures and historical behavior. When these insights are fused, Generative AI can compose structured, actionable narratives—ranging from postmortems to pre-mortems—that are not only informative but also aligned with fundamental system dynamics.

Figure 9-3 illustrates how this process unfolds within a modern observability platform. It begins with predictive AI models continuously analyzing telemetry, such as CPU trends, transaction volume, or request rates, to identify emerging patterns that signal potential service degradation. When a forecasted issue is detected, such as anticipated backend strain due to an upcoming Black Friday load spike, the platform proactively fires an alert to the SRE team. At this point, the issue has not yet impacted end users, but the system recognizes the risk early enough to act preemptively.

Next, root causal AI is invoked to interpret the surrounding context. It examines the topology, correlates metrics, and references prior incidents to identify likely root causes and propagation paths. For example, increased catalog traffic has historically led to overload in the recommendation service, ultimately causing latency in the checkout flow. This causal understanding transforms the alert from a raw signal into a diagnostic insight.

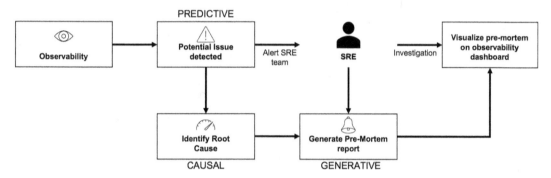

Figure 9-3. *Integrated Use of Predictive, Causal, and Generative AI to Generate Pre-mortem Reports*

Finally, Generative AI synthesizes this information into a structured pre-mortem report, which is automatically generated and attached to the observability dashboard for SRE. This report provides a high-confidence summary of the risk, the expected system behavior based on historical patterns, and recommended mitigation steps. Rather than requiring manual investigation and documentation, the SRE team sees the full context directly embedded in the service view, enabling informed action without delay. For instance, a generated pre-mortem report could look like this:

Forecasted Issue:

Projected 220% traffic increase to catalog and recommendation services may exceed capacity thresholds, risking elevated latency in the checkout flow.

Root Cause Insight:

Causal analysis links increased product search traffic to overload in the rec-service, based on similar incidents during Black Friday 2023 and 2024.

Recommended Actions:

- *Scale rec-service to 6 replicas in US-East and EU-West clusters*
- *Enable circuit breaker fallback between search and recommendation layers*
- *Set alert for checkout P95 latency > 1.5s with auto-rollout pause trigger*

Confidence Level: *High (based on matched telemetry and historical patterns)*
Owner: *Platform SRE – Retail Services*

This seamless integration of predictive, causal, and generative AI not only accelerates incident readiness but also elevates documentation from a static artifact to a dynamic, real-time asset.

While this example focuses on a pre-mortem scenario, the exact mechanism applies equally to post-incident reports, automated changelogs, periodic operational summaries, and other documentation artifacts that benefit from structured telemetry and context-aware interpretation. Generative AI is increasingly being applied across DevOps and SRE workflows to streamline documentation efforts that have traditionally been labor-intensive and inconsistently maintained. For example, incident response data—spanning alerts, logs, chat transcripts, and remediation actions—can be automatically distilled into comprehensive postmortems that include timelines, impact assessments, and contributing factors. Deployment metadata can be used to generate changelogs that highlight meaningful service updates or configuration changes, tailored to different audiences such as engineering, QA, or compliance. Architecture overviews can be synthesized from live service topology graphs, reflecting actual system composition rather than outdated diagrams. Even weekly or monthly reliability reports—often compiled manually from a patchwork of dashboards—can be generated automatically by aggregating key metrics, SLO trends, and incident patterns into a cohesive summary. In each of these cases, the common thread is context-rich, telemetry-informed documentation that evolves with the system and surfaces the information most relevant to operational decision-making.

A closely related—but more execution-focused—form of documentation is the runbook, which translates high-level knowledge into a clear, repeatable set of steps for engineers or automated systems to follow during incidents or operational tasks. This process is profoundly transformed by Generative AI. Leveraging observability-driven insights, GenAI can extract resolution patterns from past incidents, correlate them with real-time telemetry, and formalize them into structured, step-by-step procedures. This capability effectively transforms contextual insights into actionable operational assets. By utilizing structured schema formats, coupled with versioning and GitOps integration, runbooks become dynamic, automatically generated, and consistently maintained assets that evolve alongside the system, significantly reducing reliance on tribal knowledge and manual upkeep. Beyond incident mitigation, Generative AI extends to various runbook types, including generating onboarding procedures for new services, creating maintenance protocols based on historical activity, or synthesizing common sequences for repetitive operational contexts like scaling infrastructure or blue/green deployments. In essence, runbooks become living operational assets—generated, updated, and distributed automatically as an integral part of the platform itself.

Continuing the Black Friday scenario, the observability platform detects increased risk in the recommendation service and generates a pre-mortem report. From the same data, Generative AI can also produce a reusable runbook that outlines how to mitigate the issue, transforming contextual insight into an actionable asset:

Trigger:

P95 latency > 1.5 s in checkout-service traced to rec-service

Procedure:

1.   *Validate current replica count:*

kubectl get deployment rec-service -n prod

2.   *Scale service: kubectl scale deployment rec-service -n prod --replicas=6*

3.   *Monitor latency for 5 minutes via checkout-latency-p95 dashboard panel*

4.   *If no improvement, review rec-db I/O metrics and cache hit rate*

5.   *Notify on-call via #sre-alerts and attach latency snapshot*

Generated by: *GenAI—based on incident history from Nov 2023 and Nov 2024*

Owner: *Platform SRE—Retail Services*

While the previous example focused on incident mitigation, Generative AI can assist with a wide variety of runbook types beyond failure scenarios. For instance, it can generate onboarding runbooks for new services based on deployment metadata and configuration standards or create maintenance procedures by analyzing historical activity around certificate rotations, database indexing, or dependency upgrades. In repetitive operational contexts, such as scaling infrastructure during seasonal demand or executing blue/green deployments, Generative AI can synthesize common sequences into well-structured runbooks that reflect actual team practices, not just abstract best practices. As these procedures are derived directly from system activity and enriched by observability data, they evolve with the environment and reduce the reliance on tribal knowledge. In this way, runbooks become living operational assets—generated, updated, and distributed automatically as part of the platform itself.

Agentic AI for SRE

In an era of ever-increasing scale and complexity, SRE teams face the daunting task of maintaining flawless service delivery across sprawling, distributed systems. Agentic AI offers a transformative approach: autonomous, goal-oriented software agents that continuously observe your environment, learn from operational patterns, and take action to prevent or resolve issues—often before they reach human attention.

These agents augment the capacity of SRE teams to detect anomalies, forecast failures, and accelerate root-cause analysis by **blending machine learning with declarative runbooks and real-time feedback loops**. Their internal architecture includes a **cognition/brain component typically powered by LLMs**, responsible for reasoning and planning, which then enables the execution of various tasks. The sources, however, do not provide specific details on the exact platforms or programming languages (such as LangChain, Kubernetes API, Python, or Go) typically used to create these agents, nor do they mention if they are based on emerging architectures like AutoGPT or CrewAI; you may want to independently verify this information, as it is not from my sources.

By blending machine learning with declarative runbooks and real-time feedback loops, these agents become virtual SRE teammates, augmenting your capacity to detect anomalies, forecast failures, and accelerate root-cause analysis.

AI Agents Across Core SRE Practices

As site reliability engineering evolves, AI agents (Architecture: Figure 9-4) act individually (Solo Agent) or in coordinated multi-agent swarms (Swarm). AI agents can augment every dimension of our discipline, far beyond just monitoring and observability. The following explores how autonomous agents can be embedded into each core SRE practice to amplify efficiency, consistency, and innovation.

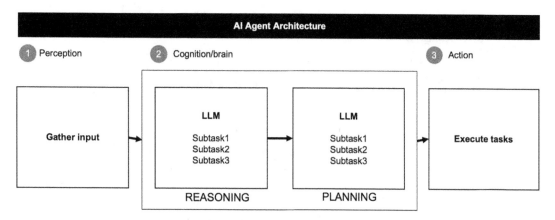

Figure 9-4. *AI Agent Example Architecture*

Incident Detection and Observability

- **Solo Agents:** Continuously parse high-cardinality telemetry, automatically tuning alert thresholds in response to usage patterns and correlating cross-service anomalies without human intervention.

- **Swarm Orchestration:** Multiple agents collaborate to triangulate incidents—one agent flags metric drift, another analyzes trace heatmaps, and a third evaluates log anomalies. For the efficient and secure operation of these autonomous agents within complex, distributed environments, it is crucial to implement robust compute and resource limits for their deployment. This ensures that no single agent or swarm consumes excessive resources, which could otherwise impact other critical systems and overall operational efficiency. Additionally, strong isolation mechanisms are paramount. These can include namespace-based throttling and fine-grained access control to limit an agent's blast radius and ensure it only interacts with authorized components, thereby maintaining system integrity and security and compliance. Such measures are vital for consistent, reliable, and reproducible infrastructure deployments and contribute to the overall resilience of the enterprise. Together, they assemble a unified incident dossier, reducing noise and prioritizing the correct root cause.

Automated Triage and Root-Cause Analysis

- **Solo Agents:** On alert, fetch relevant logs, traces, and configuration diffs; run similarity searches against past incidents; propose the top three likely failure modes.

- **Swarm Orchestration:** A team of specialized agents works in parallel. One agent assembles context, another runs dependency-graph analyses, and a third cross-references code-change metadata, converging on a high-confidence RCA in seconds.

Self-Healing and Remediation

- **Solo Agents:** Execute predefined playbooks (e.g., container restarts, auto-scale rules) the moment service health drops below SLOs, documenting each action for audit trails.

- **Swarm Orchestration:** Agents form a remediation pipeline where one agent tests a rollback in a staging slice, another verifies recovery metrics, and a third promotes the fix back to production only upon success.

Capacity Planning and Autoscaling

- **Solo Agents:** Analyze historical utilization trends and forecast resource demands, then adjust autoscaling policies to balance cost and performance.

- **Swarm Orchestration:** Different agents simulate "what-if" scenarios, traffic surges, hardware failures, and feature rollouts and collectively refine capacity thresholds and scheduling windows.

Chaos Engineering and Resilience Testing

- **Solo Agents:** Periodically inject controlled failures (network latency, CPU starvation) into non-critical environments and report on untested dependencies.

- **Swarm Orchestration:** Multiple agents coordinate a multi-vector chaos campaign: one injects latency, another disables a service node, and a third validates that circuit breakers and fallback logic behave as expected, producing a consolidated resilience score.

Change Management and Release Automation

- **Solo Agents:** Validate infrastructure-as-code diffs against security and compliance policies, gating pipelines to prevent unsafe changes.

- **Swarm Orchestration:** A group of agents emulates rollout stages: one checks policy compliance, another runs security scans, and a third executes canary deployments—automatically halting progress on any failure.

Knowledge Management and Blameless Postmortems

- **Solo Agents:** Transcribe incident call recordings, extract action items, and automatically file follow-up tickets with priority scores.

- **Swarm Orchestration:** Agents collaborate to build an enterprise-wide reliability knowledge graph; one extracts key learnings from postmortem documents, another links them to related alerts and tickets, and a third recommends documentation updates or training modules.

By embedding AI agents into these practices—whether as autonomous helpers or as tightly coordinated swarms—SRE teams can dramatically reduce toil, accelerate learning loops, and elevate reliability from a reactive function to a proactive strategic advantage.

Moreover, within this ecosystem, each AI agent takes on a specialized SRE role—whether anomaly detection, remediation execution, or resilience testing—while a master AI agent orchestrates their efforts. This master agent can pull in scripts, domain-specific knowledge, and historical incident data in seconds, automatically constructing and scaling preventive resilience measures more effectively than ever before.

Swarm SRE AI Agents Architecture

In the heat of a production situation, when, for example, the payment service begins throwing an unrelenting stream of errors, the master SRE AI agent springs into action, opening a virtual "war room" where its specialized colleagues immediately begin collaborating. This master AI agent orchestrates their efforts in a centralized manner, commissioning parallel scenario evaluations and ultimately delivering a ranked shortlist of recovery paths, each annotated with confidence scores and projected improvements in mean time to repair. This process blends machine learning with declarative runbooks and real-time feedback loops, drawing upon scripts, domain-specific knowledge, and historical incident data. Specifically, these specialized AI agents parse error spikes against recent code commits, inject harmless faults into subsystems, consult archives of compliance runbooks and throttling playbooks, and unearth past postmortems on similar incidents. This comprehensive approach allows the master agent to efficiently evaluate complex scenarios, such as testing a rollback of the latest release, simulating the activation of circuit breakers around an API, and modeling the impact of adding database replicas.

One agent quietly parses the error spikes against recent code commits, while another injects harmless faults into the billing subsystem to see if the same failures reappear. At the same time, a domain-wise SRE agent consults its archive of compliance runbooks and throttling playbooks tailored to financial services, and a knowledge-harvester agent unearths three past postmortems on similar incidents. As these insights flow in, the master agent commissions parallel scenario evaluations: testing a rollback of the latest release, simulating the activation of circuit breakers around the API, and modelling the impact of adding database replicas. Ultimately delivering a ranked shortlist of recovery paths, each annotated with confidence scores and projected improvements in mean time to repair.

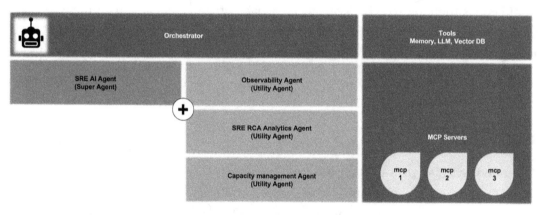

Figure 9-5. *Example of Agentic AI multi-agents Architecture*

Similarly, another production example, as a peak-traffic event looms on the horizon, the swarm assembles once more, this time to rehearse capacity and resilience in advance. One agent drives a flood of synthetic load against your service mesh, while its predictive counterpart forecasts when resource bottlenecks will emerge under different growth curves. Meanwhile, an autonomous chaos tester throttles individual compute nodes and network links to verify fallback logic, and a security-focused agent scans the latest infrastructure manifests for dangerous misconfigurations. By the end of the exercise, the SRE receives a consolidated resilience dossier that not only highlights latent weaknesses but also programs automated autoscaling rules and network adjustments, ensuring your systems stand ready when real users arrive.

This dossier, generated by a master AI agent coordinating specialized SRE AI agents, would provide a comprehensive overview of the system's readiness and recommended actions. For instance, it could detail a forecasted issue based on predictive models analyzing telemetry, such as anticipated backend strain due to an upcoming traffic spike. It would offer root cause insights derived from causal AI, linking increased catalog traffic to potential overload in specific services based on similar past incidents. The dossier would then propose recommended actions that include preconfigured automated autoscaling rules (e.g., scaling a service to a specific number of replicas) and network adjustments (e.g., enabling circuit breaker fallbacks) to mitigate the identified risks. Crucially, the dossier would feature a confidence level for its projections, based on matched telemetry and historical patterns, and outline expected improvements in Mean Time To Detect (MTTD) by proactively identifying issues before they escalate, and Mean Time To Repair (MTTR) by enabling automated remediation procedures through runbooks, accelerating resolution in record time.

In both situations, whether dousing an active outage or rehearsing for the next big surge, the multi-agent swarm acts as a seamless team of experts, transforming SRE from reactive firefighting into a proactive, predictive collaboration of autonomous SRE intelligence.

Summary

As modern systems grow in complexity and dynamism, traditional automation and monitoring approaches struggle to keep pace. This chapter explored how AI, particularly predictive, causal, and generative models, is reshaping Site Reliability Engineering by augmenting human judgment, accelerating response, and scaling operational knowledge. Predictive and causal AI enable proactive incident detection and remediation, while Generative AI is transforming how documentation and runbooks are created, maintained, and consumed. Rather than replacing engineers, these AI capabilities integrate into the SRE workflow as decision-making and knowledge-generation layers, grounded in observability data and real system behavior. For knowledge generation via GenAI, which aids in drafting postmortems, summarizing incident threads, and creating runbooks, it is crucial that this AI be grounded in a reliable data substrate to generate accurate and contextually relevant operational knowledge, thereby mitigating potential risks such as inaccuracies that might be akin to what is sometimes referred to as "hallucinations" in AI, which are not explicitly named in the sources. Similarly, predictive models that augment human judgment and accelerate incident response by surfacing hidden patterns and prioritizing signals over noise are probabilistic and can experience model drift as system behaviors evolve. To address this, these AI-driven insights need to continuously evolve with the system, requiring continuous learning from historical incidents and experiments to fine-tune the machine learning models and ensure they remain effective and reliable. Current adoption trends suggest that AI-generated insights, procedures, and documentation are poised to become standard infrastructure for reliability at scale.

## CHAPTER 10

Future and Innovations of Site Reliability Engineering

In our previous chapter, we meticulously unpacked the lessons learned, diverse use cases, and insightful case studies, offering a comprehensive view of SRE's application and the rollout of the SRE capabilities. Beyond the realms of IT, we explored its transformative impact on other projects, aiming to bolster tech and business resilience. Our discussions illuminated various innovative approaches to enhance system reliability and resilience and foster collaboration among teams even when conventional ideas seem exhausted.

Join us as we navigate through these exciting advancements, understanding how they can redefine the landscape of SRE and contribute significantly to the resilience and efficiency of technological infrastructures.

- **Evolving operational models and beyond SRE CoE**

- **Advanced patterns in resilience and scalability**

- **Building resilience in SRE with chaos experimentations**

- **Community and open-source directions**

- **Future outlook**

F. Hoeppner and F. Sbaraglia, *Mastering Site Reliability Engineering in Enterprise*,
https://doi.org/10.1007/979-8-8688-1448-8_10

Evolving Operational Models and Beyond SRE CoE

Developing a Roadmap for SRE and Chaos Engineering Implementation and Starting Your Center of Excellence (CoE)

As businesses increasingly rely on complex digital systems, the implementation of SRE and Chaos Engineering practices becomes necessary for identifying exposures before they lead to disruptive failures. In the next section, we will be developing a roadmap for SRE implementation that involves a strategic framework that guides organizations through the necessary steps, from initial planning and goal setting to execution and continuous refinement. We will learn how to establish a Center of Excellence (CoE) to get the most positive results. The CoE ensures that SRE and Chaos Engineering principles are consistently applied across the organization, facilitating a culture of resilience and continuous growth.

Beginning with SRE within a single team presents a significant challenge. Expanding it to include the entire enterprise elevates the task and the challenge to an entirely new level of complexity. The complexity increases with multiple stakeholders, with the fact that corporate politics must be understood and played well, and also, the technology complexity increases, and we must deal with competing objectives like getting things done and doing things in the right way. This section focuses on the team-level aspects, but we want to keep in mind and explain the overall roadmap. This goes hand in hand with the company-wide adoption we will share in the next section.

Our approach is characterized by agile thinking.

We want to start small, with simple and safe experiments, focusing on non-prod test environments using well-established chaos engineering tools such as **LitmusChaos, Gremlin, Steadybit, or Azure Chaos Studio**. These tools help design and run controlled failure scenarios safely. It's important to clearly define and constrain the **blast radius**, that is, the scope and potential impact of the experiment, by isolating affected services, limiting the number of impacted nodes, or setting resource thresholds, so failures stay contained and lessons can be learned without unintended consequences. Then we refine our systems, techniques, and approach based on these learnings.

The next step is to extend our impact area. We add more testing types, testing modules, and subsystems. We want to observe the effects and consider additional adjustments in our approach and system.

After that, we increase our blast radius. We want to leave the safe space with confidence and learning. We want to be in the production system with the same safety considerations because we adjusted and leaned into our steps earlier.

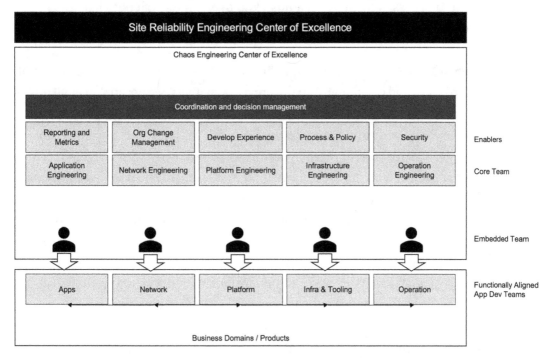

Figure 10-1. *CoE Framework and Operating Model Diagram*

This approach from a team level will be transferred to the roadmap for our company. We begin small by delving into the foundational aspects of SRE, pinpointing its significance and the value it holds for your company's leadership.

We define our first teams and run pilots. These pilots give your stakeholders confidence in Chaos Engineering within a safe space, such as a non-prod test environment. To operationalize pilots effectively, it helps to use a simple pilot rollout checklist, for example:

- Identify target systems and owners.

- Define clear blast radius and rollback plans.

- Choose tooling (e.g., LitmusChaos, Gremlin, Steadybit, Azure Chaos Studio).

- Document runbooks and learnings in a shared architecture artifact repository such as Confluence or Backstage, ensuring the knowledge stays accessible for future teams.

Through this process, we will learn, adjust, and overcome challenges like technological hurdles or conflicting goals, while identifying the most effective path forward.

At this stage of the rollout, we must start defining our Basic Center of Excellence. A small one to cover the basics and get us going. In the previous chapters, we describe the full build-out of the Center of Excellence we need to increase our impact. For the beginning, a small team of engineers doing part-time chaos engineering work will work well. The team can work in short, defined sprints on top of their normal work because we can plan the allocation in advance so that our experts can schedule their time. We want to find a shared team space where all can come together, store work items, and discuss and plan.

For our basic setup, to get the right excited people is essential. We start with the technical manager. This person has the oversight of planning and alignment and sets the objective for the pilot. We need one to two part-time chaos engineering experts to design, run, and analyze experiments. From our pilot teams, the minimum is one app developer, one infrastructure engineer, one SRE, and one person representing Operations (Production Support Team) to participate in the first experiments.

What must be accomplished by the CoE? Do we just want to increase the resilience of one team and system or roll out a resilient company?

At the start, the Center of Excellence team defines the pilot team and the resilience requirements. The nonfunctional requirements like latency, error rate, and availability for the pilot system should be documented. Then comes the next exercise, which we call "*Know your system*." The Center of Excellence must start with an inventory of systems and define what is in the scope of these systems and what is out of scope. This goes in multiple directions:

- **Know Your Software**: What modules and upstream and downstream systems are in scope?

- **Know Your People**: Contractors can be responsible for all or parts of the system. We must understand the people component, who is involved, and how we can connect with the people (management, contract, time zones, etc.).

- **Know Your Vendors**: Are the third-party vendor products in scope? It's crucial to identify how vendor systems participate in your chaos experiments. For example, you may simulate outages in vendor services by injecting failure at the integration point rather than directly disrupting their environment. Always coordinate this approach with vendors, ensuring you stay within agreed **SLA boundaries**, both legally and technically. This might include setting up joint test agreements, using sandbox environments, or agreeing on clear failure scenarios that won't breach contracts or cause unexpected impacts on shared production workloads.

- **Know Your Infrastructure**: What part of the infrastructure is in scope? Do we have a hybrid cloud environment, or is everything on our own premises? What databases, services, and dependencies are in scope? Equally important is ensuring your **observability setup** is ready to support your experiments. Confirm you have the right tools in place, such as **Prometheus**, **Dynatrace, Splunk**, or **Grafana**, to collect metrics and traces. Define the key signals you'll monitor—for example, CPU, memory, request latency, error rates, and dependency health—so you can detect issues quickly and validate that the system is behaving as expected during and after your chaos tests.

- **Know Your Business**: The application serves a business process, and sometimes multiple business partners must be identified. How are the customers in scope, and can we exclude some requirements?

The CoE team will start interacting with two groups:

- The pilot teams to test Chaos Engineering and run experiments.

- Stakeholders and leadership to report about progress and get guidance.

The objective of the Basic CoE setup is to showcase site reliability engineering and Chaos Engineering and to lay the foundation for a corporate rollout.

In closing, our exploration of the Center of Excellence (CoE) has laid the foundation for a transformative approach to resilience, emphasizing collaboration, innovation, and expertise in establishing the new SRE and Chaos Engineering Capability.

As we shift to our next section, we stand on the point of practical implementation and identification of our first chaos engineering experiment. This transition marks a key transition from theory to action, requiring us to apply our learned principles to real-world scenarios, thereby turning potential disruptions into opportunities for growth and learning.

Advanced Patterns in Resilience and Scalability

In today's hyper-connected digital landscape, building systems that are both resilient and scalable requires more than traditional testing and monitoring—it demands a proactive embrace of uncertainty. Chaos engineering is the discipline of intentionally injecting failures into your system to expose hidden vulnerabilities before they escalate into critical outages.

By simulating real-world disruptions in a controlled environment, chaos engineering enables you to identify weak spots, validate system robustness, and enhance your capacity to adapt under stress. As part of an advanced strategy in resilience and scalability, it transforms potential crises into opportunities for improvement, ensuring that your systems remain reliable, agile, and ready to scale in the face of unpredictable challenges.

What Is Chaos Engineering?

While SRE and software development teams are challenged every day by the dynamic landscape of ever-evolving technologies, where the reliability of complex distributed systems is crucial, a discipline known as **Chaos Engineering** has emerged as a strategic game changer. Site reliability engineering (SRE) is a discipline that emphasizes the operations of scalable and reliable systems, treating it through aspects of software engineering. The connection between Chaos Engineering and SRE lies in their shared goal of improving system resilience and reliability in dynamic and ever-evolving technological landscapes. This chapter serves as a comprehensive introduction to the subject, aiming to articulate a precise answer to the question: *What is Chaos Engineering? Is it really driven by random actions and chaos?*

The world of software development is an ever-evolving landscape where resilience and reliability stand as pillars of success. Enterprises have long tried to find the right balance between frequent releases of incremental product improvements and the ongoing reliability of the services they provide. Today, businesses demand high-velocity feature delivery from their tech teams, driving the widespread adoption of the Agile methodology. With **Agile**, teams reduce quarterly releases to two-week sprints and push changes faster into production. Customers want new features, and they want them quickly. But each change carries a level of risk in the production environment: more changes mean higher cumulative risk. Even if the changes are small, this can be extremely challenging for some teams, resulting in multiple releases per day. A common consequence is a higher number of incidents and costly downtimes, which threaten customer trust. One way to quantify this risk in modern CI/CD environments is with the Change Failure Rate, defined as the percentage of deployments that result in degraded service or require immediate remediation. A high change failure rate signals that your release velocity may be outpacing your system's resilience. Demand/need for rapid change in prod leads to various levels of instability and challenges "earning customer trust." Under such circumstances, a groundbreaking concept emerged: **Chaos Engineering**.

The official definition:

"Chaos engineering involves systematic and proactive experimentation on a complex system to build confidence in its capacity to tolerate challenging circumstances in production."

Examining its official definition reveals unexpected ideas, such as adopting experimentation over traditional testing, cultivating confidence, and proactively introducing failures. The SRE mindset is not to trust only the happy path but to frame experiments around the hypothesis, "What if failure/recovery scenarios?" "The goal is to identify potential failure points and correct them before they cause an actual outage or other disruption" is a key driver in "Chaos engineering." To do this safely, many teams use tools that provide automated chaos safety checks and pre-experiment validation, such as LitmusChaos Probes or Steadybit safeguards or active observability signals integrated with the Chaos Engineering tool, which verify **steady-state** conditions before injecting failure. These guardrails help ensure that experiments do not exceed the intended blast radius or impact business-critical services unexpectedly.

This methodology has revolutionized the way developers, operations teams, and **site reliability engineers (SREs)** perceive and enhance system robustness.

Now that you know what chaos engineering is, let's look at how it evolved.

Double-Click on Important Site Reliability Engineering Key Points

Before diving into the complexity of Chaos Engineering and how site reliability engineers (SREs) leverage it, it's crucial to understand foundational concepts like SLIs, SLOs, SLAs, blast radius, and hypothesis experimentation, as they are key to understanding and applying these practices effectively.

What Is an SLI?

A Service Level Indicator (SLI) is a specific metric used to measure the performance of a service. It provides a quantifiable indicator of some aspect of the service's operation, such as service availability, request latency, error rate, service saturation, or service throughput.

What Is an SLO?

A Service Level Objective (SLO) is a target value or range for an SLI that a service provider aims to achieve over a specific period. It represents the desired level of service reliability and performance from the customer's perspective. For example, an SLO might be to ensure that 99.9% of all requests are processed within 200 milliseconds in the last period of 28 days. SLOs are critical for setting and maintaining expectations between service providers and users, ensuring that the service meets or exceeds what users consider acceptable. Example OpenSLO YAML is shown in Listing 10-1.

Listing 10-1. OpenSLO YAML definition example

```
### YAML
apiVersion: openslo/v1
kind: SLO
metadata:
  name: sloth-slo-checkout-service
  displayName: Requests Availability
spec:
  service: my-service
```

```
description: "Common SLO based on availability for HTTP request
responses."
budgetingMethod: Occurrences
objectives:
  - ratioMetrics:
      good:
        source: prometheus
        queryType: promql
        query: sum(rate(http_request_duration_seconds_
        count{job="checkout-service",code!~"(5..|429)"}[{{.window}}]))
      total:
        source: prometheus
        queryType: promql
        query: sum(rate(http_request_duration_seconds_count
        {job="checkout-service"}[{{.window}}]))
    target: 0.999
timeWindows:
  - count: 30
    unit: Day
```

What Is an SLA?

A service level agreement (SLA) is the "promise" and the formal agreement between a service provider and a customer that defines the expected level of service, often including the consequences if the agreed-upon service levels are not met, such as financial penalties or credits. SLAs are legally binding documents that outline the responsibilities of both parties and provide a framework for managing service performance and expectations.

What Is a Blast Radius?

In the context of Chaos Engineering, the blast radius refers to the scope or extent of impact that a failure or disruption can cause within a system. It defines all boundaries of the system or components that can be affected by a chaos experiment. Managing the blast radius is crucial to ensure that chaos experiments do not inadvertently cause widespread damage or disruption to critical services. The goal is to limit the impact to a controlled, manageable area while still gaining valuable insights into the system's resilience.

What Is a Hypothesis?

A hypothesis is a statement or theory about how a system will behave under specific conditions of stress or failure introduced during a chaos experiment. It forms the basis of the experiment, guiding the design and execution of the test. The hypothesis typically asserts that the system will continue to operate normally or that a particular service will degrade in a predictable manner. After the experiment, the results are analyzed to determine whether the hypothesis was correct, helping to identify potential weaknesses or confirm system robustness. Example: *If we inject additional traffic that increases network latency by 200 ms, then our application's 95th percentile response time should remain under 1 second.*

Chaos Engineering (CE) and site reliability engineering (SRE) are distinct yet complementary disciplines: while SRE focuses on ensuring service reliability, automation, and operational efficiency, Chaos Engineering involves intentionally introducing failures to "test" system resilience. Interestingly, SREs are the primary advocates for Chaos Engineering, often adopting it in the more advanced stages of the SRE maturity lifecycle, typically **after SRE foundational practice adoptions like SLO definition, automation, incident response workflows, and observability are in place**. This ensures teams have clear baselines, visibility, and steady-state metrics before experimenting with controlled failures. Paradoxically, Chaos Engineering would not exist without the foundational principles of SRE, as it became apparent from the need to continuously improve system reliability and robustness.

History of Chaos Engineering

Chaos Engineering finds its roots in the innovative ecosystem of **Netflix**. In 2010, the team at Netflix faced a key challenge—maintaining a flawless streaming service in an increasingly complex, cloud-based environment. Understanding the significance and importance of the core streaming system to Netflix is essential. Without the streaming service, there was no Netflix. People would sit in front of the television on a Saturday night with popcorn but without entertainment from Netflix. Imagine how sad and, in fact, against Netflix's core business. To address this, they had the idea to develop an unexplored approach—intentionally introducing controlled chaos into their systems. The notion was to proactively inject weaknesses, exposures, and failure points within the architecture to fortify it against unexpected outages or failures. This experimental mindset led to the birth of *Chaos Monkey*, an automated tool designed to randomly terminate production instances within Netflix's infrastructure. The controlled chaos served as a mechanism to enhance system resilience, with learnings from these disruptions leading to more robust, more reliable systems. Even if this concept was developed to guardrail the most significant applications and not each company is Netflix, the concept has evolved. The concept of Chaos Engineering is now widespread and sees mainstream adoption. Having explored the historical evolution of chaos engineering, this allows us to shift our focus to explore the essential benefits that this discipline brings to the domain of system reliability and resilience.

Key Benefits and Impact of Chaos Engineering in IT Operations and Delivery

IT organizations are no longer cost centers. IT makes up the fabric of the entire enterprise. Its mission is speed. Something in IT has changed in the last ten years. The business department requests high speed from their IT delivery because this is in fact how they beat the competitors and how they make money. Developers must deploy frequently in production, often several times per day (**Deployment Frequency**) in short iterations, eventually from code commit to production deployment in less than one day (**Lead Time for Changes**). These benchmarks help teams track how quickly they can deliver reliable changes and adapt to customer needs.

Current business strategy requires acceleration from IT in small and fast changes. It requires small innovations and quick pilots to test and try out new ideas. This is the new way of interacting with customers. Customers are demanding fast improvements and ongoing changes. We, as end-users, are used to expecting new features daily in short iterations, waiting for the next phone OS upgrade, smartwatch update, or our favorite app's new release with a bunch of cool features. This is what they know from their App Store. My sports have been improving all the time. My car gets regular updates and improves its capabilities. This is what we expect and what we are used to.

More change can be seen in how money is spent on IT. In this area, two souls live in our breast: building and operations. In most companies, we see too much spent on operations and too little on engineering. For each dollar companies spend on engineering, we have seen that they spend double or more on operations.

Each area is dependent on and connected to the other. Changes in production and human intervention in operations are the main cause of the growing number of incidents. Missing automation in operations leads to more manual work and causes even more disruptions in production systems. As complexity increases, so does the time to find errors and provide solutions.

In traditional companies, the pain caused by incidents is poorly distributed. The symptoms lie with the customer and the production support. The cause lies with the developers. The engineer who created the code does not feel the outages. They don't get notices when their code causes an interruption. Developers do not even see the budget spent on operations. The budget is discussed only on a leadership level. Engineers normally don't know the cost of their application in operation, the number of incidents, and the service requests. Most do not even know the names of the operations experts.

Imagine if the developer team were accountable for the operational spending. They would have to buy the operations service or even spend their own time doing the manual operation tasks. They would have to fix incidents and buy cloud storage. Even better, imagine they would also experience the positive effects of their application. Imagine they could see the rising numbers of customers loving their work and the positive resonance a feature has on the customers. And finally, imagine they would even get a part of the earnings from the use of the application. What would this world look like?

Resilience has become a competitive advantage for companies. Companies that go above the required regulatory requirements stand out. Imagine a car manufacturer making a commercial that points out that their autopilot has fewer accidents than their competitor's system. Or imagine a trading platform communicating that their application uptime is the highest in the branch.

Summarizing what we just described:

- IT processes are speeding up more and more but lack safety guidelines. We are driving our car fast but without seatbelts. We've sped up our deployments, but we aren't considering additional stability improvements.

- The developers are becoming queens and kings. It is the core of every product and customer experience. It determines the reputation of a firm, and it will do so even more with the upcoming Metaverse.

- Spending on operations is out of control and increasingly is not addressing the tasks that are actually causing the costs.

By working together with enterprises on their SRE and Chaos Engineering transformation, we have realized that resilient companies share some core beliefs:

- Metrics and data are the fundamentals.

- Build guardrails to deliver on the trust model across the entire system architecture and improve the system; don't rely on a single person.

- Promote and influence a "proactive" versus "reactive" mindset.

- Build a culture to "adapt" for change; all and everything is changeable.

- Access each change and complexity; strive for simplicity.

- Improve collaboration; share your knowledge always, anytime.

- Be prepared and confident in your ability to fix the problem in a timely manner (with predefined SLAs). Develop, test and deploy adhering to SLAs (service level definition).

After highlighting the benefits of chaos engineering, let's zoom in on its vital contribution to enhancing system reliability—indispensable for domains such as site reliability engineering (SRE), DevOps, and Ops teams.

Importance of Chaos Engineering in SRE, DevOps, and Operations Teams

Chaos Engineering revolutionizes the traditional paradigm of system reliability by fostering a proactive rather than reactive approach. DevOps and SRE practices intrinsically aim to create and maintain highly resilient, stable, and scalable systems. Chaos Engineering complements these goals by providing a structured methodology to

- **Proactively Identify Weaknesses**: By simulating real-world failures in a controlled environment, weaknesses and exposures within the system can be uncovered and fixed before they impact end-users.

- **Build Resilience**: Understanding and mitigating system vulnerabilities through controlled chaos enables the construction of more resilient architectures, reducing downtime and enhancing user experience.

- **Enable Continuous Improvement**: Chaos Engineering infuses a culture of continuous improvement by promoting the DevOps mindset, believing that all systems can fail and that failures are seen as opportunities for learning and building systems that are resilient to failures.

- **Enhance Collaboration**: It encourages cross-team collaboration, where developers, operations, and SREs work together to build more reliable systems, fostering a shared responsibility toward system reliability.

- **Build Confidence**: It is important to build trust and the belief that the system is correctly designed and can withstand any failure conditions; it can self-heal and restore functionality. This will help proactively to minimize "blast radius" for outages and help SREs to quickly identify, isolate, and mitigate system downtime.

The ultimate goal of Chaos Engineering within the DevOps and SRE domains is to create systems that not only withstand failures but also continue to deliver exceptional and expected performance and reduce "Unknowns." When correctly applied, Chaos Engineering contributes to more robust and reliable systems. In an enterprise context, this means fewer incidents and outages. These are improvements the business department and the customer realize and appreciate. The main dimension Chaos Engineering teams are improving in the context of SRE and DevOps is testing resilience to identify vulnerabilities proactively. When we learn about a system's behavior, we can implement improvements to increase fault tolerance—the ability of the system to withstand and recover from failures. The concept allows teams to simulate real-world issues in a controlled environment. By doing so, we uncover weak points. The final goal is *congruent with SRE: to ensure a seamless and positive user experience even when facing system failures.*

As we progress through this book, we will explore the core principles, best practices, and real-world applications of Chaos Engineering, unraveling its potential to transform the landscape of system reliability and resilience in the ever-dynamic world of technology.

Chaos Engineering Experimentation Stages: Building Resilience through Chaos Engineering

As the narrative of system reliability unfolds, we reach a pivotal milestone where theoretical knowledge transitions into practical application. This chapter presents a detailed roadmap through seven critical phases, from initial ideation to real-world application, ensuring that when the unexpected strikes, the systems stand ready, resilient, and responsive.

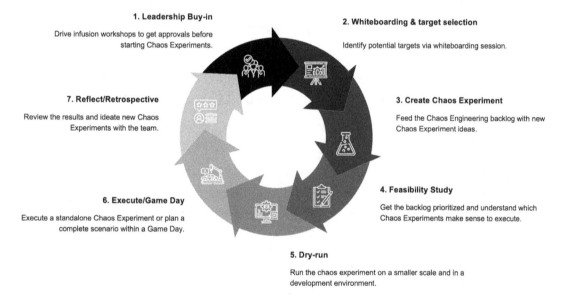

Figure 10-2. *Diagram of Chaos Engineering Experimentation Lifecycle*

Here is an explanation of each step in the flow process:

1. **Get the Buy-In of Leadership**: Leadership, executives, and business owners have to familiarize themselves with the concept of chaos experimentation to avoid surprises and make it transparent at all stages.

2. **Whiteboarding and Target Selection**: This is the brainstorming phase where potential targets for chaos experiments are identified. These could be parts of the system that have recently had issues, are poorly understood, or have been typically avoided by the team.

3. **Creating Chaos Experiments/Scenarios**: Once targets are identified, specific chaos experiments or scenarios are defined. These are designed to simulate potential disruptions to the system. Each scenario must have a plan for rolling back the changes in case something goes wrong (abort plan).

4. **Feasibility Study**: This step involves deciding which of the chaos experiments or scenarios will be most effective for the target in question. The decision is based on a feasibility check and an evaluation of the risk potential associated with the experiment.

5. **Dry Run**: Before fully implementing the chaos experiment on a smaller scale or using a dedicated development environment, a dry run is performed to ensure that it works as expected. This step also involves confirming the criteria for when to abort the experiment and how to roll back the changes.

6. **Execute/Game Day**: This is the execution phase where the chosen chaos experiments or scenarios are carried out. The team observes the system's behavior and takes note of the results.

7. **Reflect/Retrospective**: After the chaos experiments are completed, the team reviews the results. They create bug reports or change requests for the backlog for any issues discovered during the experiments. Surprising findings or unidentified issues are noted and may become targets for future chaos experiments.

The purpose of this process is to proactively find and fix problems before they affect users, ensuring that the Chaos Experiment is successfully executed without collateral events besides the Chaos Experiment.

Get the Buy-In of Leadership

Getting the buy-in of leadership, executives, and business owners for chaos experiments is required because these experiments involve intentionally injecting failures into systems to stretch resilience and reliability. Such actions carry inherent risks and require resources, which can be justified only if the leadership understands and supports the potential benefits.

Why Leadership Buy-In Is Important:

1. **Resource Allocation**: Chaos experiments can require significant investment in terms of time, personnel, and technology. Executives and business owners must approve these resource allocations.

2. **Risk Management**: Leaders are responsible for managing risk. Since chaos experiments can potentially disrupt services, executives and business owners must understand and agree to the level of risk being taken.

3. **Cultural Change**: Chaos Engineering often requires a cultural shift toward accepting failures as a path to improvement. Executives and business owners set the tone for organizational culture and can champion this change.

4. **Strategic Alignment**: Experiments should align with the company's strategic goals. Leadership can ensure that chaos engineering initiatives contribute to broader business objectives.

5. **Sustainability**: For chaos engineering to be a sustainable practice, it needs to be integrated into the regular workflow. Leadership support is crucial to make it an ongoing practice rather than a one-off experiment.

How to Get Leadership Buy-In:

1. **Educate About Benefits**: Clearly articulate how chaos experiments can improve system resilience, save costs in the long run by preventing more significant outages, improve customer satisfaction by providing a more reliable service, validate SLOs, and enhance the expertise of SRE teams.

2. **Present Case Studies**: Share success stories from other companies that have successfully implemented chaos engineering. Quantifiable results from respected peers can be particularly persuasive.

3. **Align with Business Objectives**: Show how chaos experiments support the organization's goals. For example, if the business aims to be the most reliable service provider in the market, chaos experiments are directly aligned with that goal.

4. **Start Small**: Propose a small-scale experiment with limited scope and risk. A successful small experiment can demonstrate the value of chaos engineering and lead to support for more extensive testing.

5. **Risk Assessment**: Accompany your proposal with a thorough risk assessment and a mitigation plan. This demonstrates responsibility and preparedness.

6. **Define Metrics for Success**: Establish clear metrics that will be used to measure the success of chaos experiments. This could be the number of vulnerabilities discovered, uptime improvement, or reduced incident response time.

7. **Create a Roadmap**: Develop a roadmap that outlines the steps, timeline, and expected outcomes of the chaos experiments. This helps leaders see where the initiative is going and how it fits into the larger picture.

8. **Engage in Dialogue**: Be open to questions and concerns from leadership. Engaging in a two-way conversation can help address any hesitations and refine the approach to chaos experiments.

9. **Demonstrate Expertise**: Show that the team conducting the experiments has the necessary expertise and that external experts are involved if needed. This reassures leadership that the experiments will be conducted responsibly.

By addressing the benefits, aligning with business objectives, and demonstrating a responsible approach to risk, you can effectively secure leadership buy-in for chaos experiments.

Whiteboarding and Target Selection

It is the very first stage of gathering information and requirements for the Chaos Experiment. It will be delivered in the form of multiple workshops; sometimes one is enough, but for complex scenarios, it is necessary to plan multiple workshops.

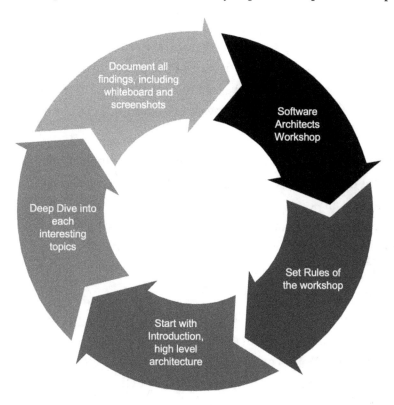

Figure 10-3. *Whiteboard Cycle*

The essential tool here is the "whiteboard," digital or physical, and we usually consider the following steps:

1. **Consider Getting the Software Architects in One Meeting:** This step emphasizes the importance of bringing multiple personas together, all the key technical stakeholders, for an open and blameless technical discussion.

2. **Clearly Set the Rules of the Workshop, and Talk Openly About Potential Weaknesses, Typical Mistakes, and Typical Implementation Issues:** Encourages an open blameless discussion about possible exposures within the architecture, common errors that could occur, and issues generally encountered during the implementation phase.

3. **The Workshop Starts with an Introduction of the Architecture on a High Level and Deep Dive into Each Interesting Topic:** The aim here is to begin the conversation focusing on the overarching design and structure of the software or system without getting into too much detail.

4. **Protocol All Findings in a Document with Screenshots or Photos of the Architecture and Whiteboard:** It is necessary to document all points of the discussion in a central shared portal (e.g., confluence), marking the facts that are creating tension in the team, including postmortem of past incidents. The workshop can also be planned to use a previous analysis of past incidents; this will give a better direction to the workshop.

5. **Repeat the Workshop Until:** All topics are clarified, ensure that all parties fully understand all aspects of the discussion, and identify at least three "**Points of interest**." Based on our experience, this seems to be a metric for the meeting's productivity.

Definition of "Points of Interest":
In the context of defining "Points of interest" within a Whiteboarding's workshop stage, the term refers to specific areas or scenarios that require special attention. A "Point of interest" arises when there is uncertainty about how the application will react under certain conditions, signaling a need for more detailed understanding and experimentation. It also emerges when the architects and the team hold differing opinions on how to approach a particular scenario, underscoring the need for further discussion to reach a consensus. When there is a varied interpretation of the requirements among team members, it highlights a gap in understanding that necessitates alignment. Additionally, when architects collectively determine that certain aspects of the architecture warrant a closer look, it is a mutual recognition that those areas should be subject to a more in-depth review to ensure robustness and correctness.

To recapitulate, follow some examples of "Points of interests":

- Nobody is sure how the application will react.

- The architects have different opinions on a scenario.

- Different understanding of the requirement.

- The architects concluded that this should be checked in detail.

Creating Chaos Experiments/Scenarios

This stage is meticulously prepared to ensure a structured approach to the Chaos Experiment.

Table 10-1. *Chaos Experiment Checklist Template*

Chaos Experiment Data	Description	Input
Identification of the target system	Pinpoint the system component for experimentation	
Hypothesis formation	Draft an initial hypothesis with "what if" questions to predict potential system responses and behavior	
Selection of the experiment type	Choose an experiment type, ranging from standard scenarios to custom attacks	
Crafting the chaos experiment scenario	Outline a detailed scenario of the chaos experiment from start to finish	
System load consideration	Decide on simulating typical system load during the experiment for accuracy	
Abort plan formulation	Develop a clear plan to revert changes and stabilize the system if necessary	
Monitoring and metrics	Establish "golden signals" for monitoring and health checks to protect system integrity	
Determining duration and frequency	Decide the duration and frequency of experiments to optimize learning and operational stability	

Here is an enhanced explanation of each component and consideration:

- **Identification of the Target System**: Begin by pinpointing the system component that will undergo the experimentation. This sets the stage for all subsequent steps.

- **Hypothesis Formation**: Draft an initial hypothesis by asking "what if" questions to anticipate potential system responses to the chaos introduced. This forms the backbone of your experiment's rationale.

- **Selection of the Experiment Type**: Choose the most fitting experiment from a spectrum that includes standard, out-of-the-box scenarios to more tailored, custom attacks, each designed to probe different facets of your system's resilience.

- **Crafting the Chaos Experiment Scenario**: Outline a comprehensive scenario that details each step of the chaos experiment, from initiation to conclusion, ensuring a clear narrative of the intended disruption and its purpose.

- **System Load Consideration**: Determine if there is a need to simulate typical system load (e.g., Locust, JMeter, etc.) during the experiment to reflect operational conditions more accurately.

- **Abort Plan Formulation**: Develop a well-defined abort plan to revert any changes and stabilize the system swiftly should the experiment lead to unexpected or hazardous states.

- **Monitoring and Metrics**: Establish critical "golden signals" and implement monitoring health checks that will trigger the cessation of the experiment if certain thresholds are breached, thereby safeguarding system integrity.

- **Determining Duration and Frequency**: Decide on the length of each experiment and the intervals at which they are conducted to balance the depth of learning with the system's operational demands.

Feasibility Study

In essence, a feasibility study is a critical step in chaos engineering that ensures experiments are safe, effective, and aligned with organizational goals. It helps teams make informed decisions and prepare adequately, ensuring that the benefits outweigh the risks.

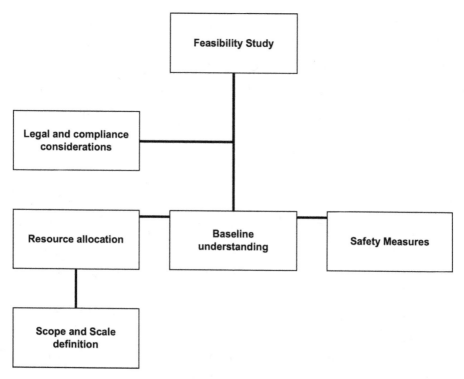

Figure 10-4. *Feasibility Study Decision Tree*

The feasibility study has five important reasons:

- **Risk Assessment:** It helps identify potential risks and impacts of chaos experiments, enabling informed decision-making and preparation for unintended consequences.

- **Resource Allocation:** Determines the resources needed for chaos experiments, ensuring that the investment in time, personnel, and technology is justified and well-planned. Not only for the execution of the Chaos Experiment, but also to recover from unexpected conditions.

- **Scope and Scale Definition:** Clearly defines which parts of the system will be experimented on and to what extent, ensuring focused and manageable experiments.

- **Baseline Understanding:** Establishes baseline metrics and performance indicators crucial for measuring the impact of the experiments and understanding normal system behavior.

- **Legal and Compliance Considerations:** Identifies any regulatory issues (e.g., EU-GDPR) that need to be addressed, ensuring that experiments comply with legal requirements.

- **Safety Measures:** Plans for safety measures and fail-safes to quickly restore the system in case of unexpected problems, ensuring minimal disruption.

Dry Run

Conducting a short dry run of chaos experiments in a development or dedicated environment is essential before running them in production to ensure controlled and safe testing. It is not about actually to run the experiment, but about confirming wether the chosen scripts and target can execute. For example, if there is proper execution permission (RBAC), permission to stop the experiment, if technical user is able to read logs and metrics, and simply if we have access to the specific observability platform used for the system. This preliminary step helps identify unexpected behaviors and validates the design of the experiments, ensuring they behave as expected and that the tools and failure injections are accurate. It significantly minimizes risks, such as permission and monitoring issues. Additionally, it allows teams to train and prepare, understanding their roles and responses during experimentation. Moreover, it ensures that the experiments do not inadvertently violate any compliance requirements or security policies, avoiding potential legal and security issues.

Here are some focus areas to keep an eye on during the dry run:

- **Observability:** Ensuring the system's metrics are properly monitored and can catch the system's steady state.

- **Expected Resilient Architecture:** Checking the system's redundancy and fault tolerance, including servers, databases, autoscaling, and load balancing.

- **Hypothesis and Expectations**: Being confident that the system will behave as expected during the tests.

- **Alignment to Respond**: Preparing for incident and problem management, ensuring all alerts are set, and conducting postmortems.

- **Reversibility**: Having a clear plan to revert changes if necessary, including a "big 'Stop' button."

Execute/Game Day

Best practice is to execute the Chaos Experiments standalone or as part of a game day.

In the **execution phase** of a chaos experiment, the planned disruption is actively introduced into the system in production. This is where the theoretical meets reality, as the team implements the fault injections or shutdowns designed during the planning phase. The system's response is closely monitored to gather data on performance, resilience, and failure recovery. This phase is critical for observing the actual impact of the chaos on the system, understanding its weaknesses, and identifying areas for improvement.

A **game day** is an event where the team comes together to conduct, observe, and analyze the chaos experiment in real time.

To run a game day safely and effectively, always have a simple checklist:

- Confirm clear objectives and hypotheses.

- Define the blast radius and rollback steps.

- Set up observability dashboards to monitor steady-state.

- Communicate an escalation plan so everyone knows who to call if things go wrong.

As a best practice, avoid scheduling game days, or any major deployments, on Fridays, as recovery windows may be limited heading into weekends.

It is a collaborative effort involving engineers, developers, and sometimes even business stakeholders. During this session, the team executes the planned chaos scenarios, monitors the system's response, and practices the incident response strategies. The game day is not just about breaking things; it is a focused learning exercise to improve system robustness and team readiness. It is a day of controlled chaos with the goal of building confidence in the system's resilience and the team's ability to handle unexpected disruptions.

Reflect/Retrospective

The reflect and retrospective phase is a crucial iterative step in the agile process of chaos engineering, where the team convenes after the experiment to analyze outcomes and share learning collaboratively. This phase is not the conclusion but rather a vital moment for continuous improvement and innovation. The team reflects on the collected data from the experiment to understand the successes, failures, and root causes. They engage in a dynamic discussion about unexpected system behaviors and the efficacy of their response strategies, promoting an agile mindset of adaptability and resilience. This reflective dialogue is instrumental in identifying opportunities for enhancement in both the system's robustness and the team's agility in incident management. It is an environment that encourages open feedback and collective problem-solving, integral to fostering an agile culture of ongoing learning and evolution.

Example 4Ls Retrospective Template:

- **Liked**: "We liked how quickly we resolved incidents during the chaos experiment."

- **Learned**: "We learned that our alerting thresholds were too narrow and caused noise."

- **Lacked**: "We lacked clear ownership for rollback steps."

- **Longed For**: "We longed for better observability dashboards to see real-time impacts."

The insights and ideas generated here become the sources for new, more targeted chaos experiments, keeping the cycle of improvement animated and ongoing. In essence, this phase transforms the experience into innovation, constantly propelling the team forward in their agile journey of chaos experimentation.

Get Started with Chaos Engineering Experiments

Chaos engineering borrows its core methodology from the scientific method, which is grounded in the continuous cycle of formulating hypotheses, conducting experiments, observing outcomes, and then iterating on the process. Broadly, like in scientific research, chaos engineers start by forming a hypothesis that predicts how a system will behave under certain conditions. This hypothesis often takes the form of a "what if" question, such as "**What if this service unexpectedly becomes unavailable?**"

Following the formulation of a hypothesis, chaos engineers design and execute experiments to test their beliefs, carefully introducing variables—in this case, disruptions or "chaos"—into the system in a controlled way. These experiments are akin to scientific trials where conditions are observed, data is collected, and outcomes are analyzed.

The observation phase is where engineers monitor the system's performance against the expected behavior drafted in the hypothesis. They use various metrics and monitoring tools to evaluate the impact of the chaos introduced and determine whether **the system behaves as hypothesized.**

Once the experiment is concluded and observations are made, engineers analyze the results, often revealing more insights about the system's resilience and areas for improvement. This analysis may **confirm** the initial hypothesis, **decline** it, or **expose unexpected behaviors**. Based on these findings, engineers can then refine their hypotheses or form new ones, leading to further experiments. *Not all experiments are successful; in fact, the term "successful" is matter-of-fact for what is willing to be proved.*

This iterative cycle—hypothesize, experiment, observe, analyze—is repeated continuously, driving improvements in system reliability and performance. Each iteration builds upon the learnings of the previous one, creating a loop of continuous improvement that is common to both scientific exploration and chaos engineering.

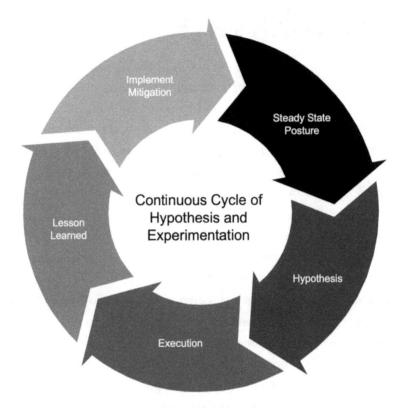

Figure 10-5. *Continuous Cycle of Hypothesis and Experimentation*

Some of our best practices and examples:

- Monitor and write down which metrics indicate the system is in a steady state.

- Visualize critical user journeys to understand the impact of experiments.

- Anticipate unknown behaviors by listing all possible ones.

- Verify redundancy and autoscaling capabilities to ensure they're working as expected.

- Confirm that fault tolerance mechanisms are in place.

- Ensure there is confidence that the system will perform as hypothesized and avoid testing on an already broken system.

- Document the expected incident and problem management process in detail.

- Make sure all alerts are configured and ready to notify of any issues.

- Conduct postmortems for experiments to learn from the outcomes.

- Have a prominent and accessible "Stop" button to halt the experiment if needed.

- Always have a plan ready to revert the system back to its original state.

Create the First Hypothesis

Creating the first hypothesis in chaos engineering implicates framing a knowledgeable guess about how a system will respond under specific conditions of pressure or failure. This initial hypothesis is based on the current understanding of the system's design, architecture, and behavior. It is a predictive statement that sets the **expectation** for what will happen when the system is subjected to a chaos experiment.

Here is how to create this first hypothesis:

1. **Understand the System:** Start with a detailed and careful understanding of the system's normal behavior, including its architecture, dependencies, and how it handles transactions and failures under regular conditions. It does not need to be only when something is going wrong but also considers when everything works correctly, and, for example, the system needs to scale up or down.

2. **Identify Potential Failure Points:** Look for components or services within the system that seem most weak or critical to overall operations. These could be areas with complex logic, known bugs, or high load.

3. **Consider the Worst-Case Scenarios:** Ask "what if" questions that challenge the system's resilience. For example, "What if the network latency between the backend and database suddenly increases?", "What if the database suddenly loses connectivity?", "What if there is a sudden spike in traffic?", or "What if the load balancer has a network issue?"

4. **Use Past Incidents:** Reflect on past outages or incidents for hints about where the system might carry weaknesses that have not been fully addressed.

5. **Predict the Outcome:** Based on the understanding of the system, predict what the belief is and what happens when we introduce a specific kind of controlled disruption. For instance, "If the database loses connectivity to the backend, then the backend can still communicate with the caching layer that will handle requests until the day

The key idea is that in the process of formulating hypotheses, there's no definitive right or wrong. It's a cyclical process. Our objective isn't to prove our beliefs as true or false. Rather, each hypothesis serves as a tool for deepening our understanding of our system.

After generating multiple hypotheses about how a system might behave under different conditions, these hypotheses are organized into a classic backlog, much like user stories in an agile scrum framework. This chaos experimentation backlog serves as a list of possible experiments to be organized, validated, and conducted, allowing teams to systematically work through each validated hypothesis and assess the system's resilience.

Here is an example of a user story for a chaos experiment:

User Story: 1—Chaos Experiment Load Balancer Connectivity

Title: Load Balancer Connectivity Disruption Experiment

As a Site Reliability Engineer

Target building blocks Infrastructure

I want to simulate the loss of connectivity to our primary load balancer

So that I can verify that the traffic is automatically and efficiently rerouted to secondary load balancers without service disruption.

Acceptance Criteria:

1. The primary load balancer is intentionally disconnected during a controlled experiment.

2. Traffic is immediately and seamlessly redirected to the secondary load balancers.

3. There is no noticeable impact on the user experience, with all requests being served as expected.

4. The auto-scaling system responds appropriately to the change in traffic through secondary load balancers.

5. System resilience is maintained, and performance metrics such as response time, throughput, and error rates remain within predefined acceptable limits.

6. The incident is logged, and the monitoring system alerts the team within 1 minute of the occurrence.

7. A fallback mechanism is triggered, ensuring service continuity, and a postmortem analysis is conducted to document lessons learned and areas for improvement.

This user story emphasizes the need for a resilient **infrastructure** that can handle unexpected load balancer and connectivity failures, ensuring high availability and reliability of the service. The hypothesis should target only one building block at a time; more complex Chaos Experimentation scenarios are in fact a chaining of multiple Chaos Experiments, executed in a defined time schedule.

Community and Open-Source Directions

While this book has covered a broad range of topics, several areas still require further exploration, such as the impact of site reliability engineering (SRE) on security practices and the integration of GenAI (e.g., CrewAI, LangChain, etc.) and Agentic AI. Additionally, questions about the ethical implications of balancing system reliability with operational trade-offs remain open for discussion. These unanswered questions present opportunities for future research and development in the field.

The principles of SRE align with broader industry trends towards continuous improvement, agile methodologies, and a culture of innovation, fitting seamlessly into concepts like Agile, DevOps, and platform engineering. As organizations strive to deliver more reliable and resilient services, the adoption of SRE practices will become increasingly crucial. This shift reflects a growing recognition of the need to design and operate systems that are not only stable but also capable of scaling and recovering quickly from unexpected disruptions.

At the beginning of this book, we posed the question, "How can we ensure our systems are reliable and resilient under ever-evolving conditions?" Through the exploration of SRE, we have provided a framework for answering this question. By systematically applying SRE principles such as error budgets, service-level objectives (SLOs), and continuous monitoring, we can build systems that not only endure but also adapt to the uncertainties of real-world operations. From building systems, we can take the next step to establish robust platforms and products that scale with confidence, rooted in a culture of resilience and continuous learning.

Throughout the book, we have highlighted various case studies. These real-world examples illustrate the practical application of SRE principles and the tangible benefits achieved. By studying these cases, readers can gain insights into how to implement similar practices within their organizations.

Implementing SRE has its challenges, including cultural resistance, the complexity of defining meaningful SLOs, and the intricacies of balancing reliability with innovation. However, these challenges also present opportunities for growth, learning, and innovation. By addressing these obstacles head-on, organizations can develop more resilient systems and foster an adaptive engineering culture.

To continue your learning journey, consider conducting a reliability review or a blameless postmortem within your organization and documenting the results. Reflect on the insights gained and share your findings with your team or the broader SRE community. Additionally, ponder questions such as "What are the most critical components of our system that require enhanced reliability?" and How can we foster a culture that views incidents as opportunities for improvement?"

In future research in SRE, we will explore the integration of machine learning to predict and automate incident responses, as well as the development of more sophisticated metrics to assess system health and reliability. Additionally, investigating the impact of SRE on organizational culture and cross-functional collaboration could provide valuable insights. These areas of study will help advance the field and uncover new ways to enhance system reliability.

To apply the knowledge gained from this book, start by identifying key components of your system that would benefit from improved reliability practices. Focus on applications where stability is critical and failure is not an option. Develop a plan for implementing SRE principles such as error budgets and SLOs, gradually scaling their application as you build confidence and expertise. Engage with your team to foster a culture of reliability and continuous improvement, using the principles of SRE to drive better outcomes for your organization.

As we conclude this chapter, we delve into understanding the specific areas where contributions from site reliability engineering (SRE) practitioners are most needed. This knowledge helps in pinpointing where efforts should be concentrated to maximize the impact of SRE practices. By identifying these critical areas, organizations can direct their resources and innovation efforts more effectively, ensuring that they tackle the most pressing challenges with precision and strategic focus.

Under the section Areas for Further Research and Development in SRE, we identified exciting avenues for future exploration. This includes the development of more sophisticated tools and methodologies that can predict system failures with greater accuracy, as well as the potential for personalized SRE frameworks tailored to the unique needs of different industries. Horizontal and vertical SRE agents can play a pivotal role in driving these innovations by leveraging their specialized expertise and collaborative insights.

We predict the trend of horizontal SRE agents and vertical agents as key components of a robust, future-ready, scalable SRE strategy. Horizontal SRE agents focus on cross-cutting concerns such as alignment with business outcomes and orchestrating vertical agents. Vertical agents, on the other hand, specialize in addressing domain-specific reliability challenges, tailoring solutions to the unique needs of the specific capability. This dual approach ensures both breadth and depth in addressing reliability concerns, creating a balanced and holistic framework for operational excellence.

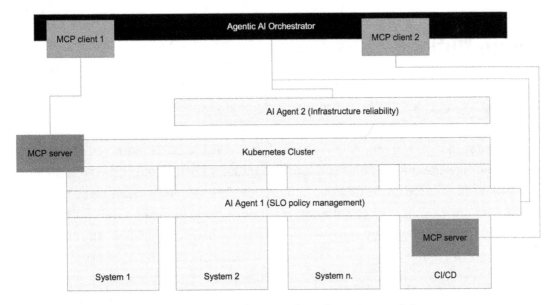

Figure 10-6. *Agentic AI Example of Vertical and Horizontal Agents*

Furthermore, we explored emerging trends in AIOps and how these will shape the future of SRE. In both directions, AIOps can benefit from SRE practices, and vice versa. The integration of AIOps with SRE promises a transformative shift in how organizations automate complex decision-making processes, enhance real-time analytics, and anticipate system behaviors under stress. This convergence is poised to dramatically refine the scope and effectiveness of SRE strategies, enabling teams to proactively identify and address potential reliability issues before they impact end users.

Finally, in our final reflection on SRE, we considered the broader implications of these developments. We acknowledged the continuous evolution of SRE as a discipline that not only safeguards systems but also nurtures an environment where resilience is continuously enhanced and valued. By fostering a culture of reliability, organizations can ensure that SRE practices remain aligned with broader business goals and priorities.

As this chapter closes, consider how these skills, insights, and future directions can be applied not only to address current technical challenges but also to pave the way for future innovations in monitoring, system reliability, and overall business resilience. The journey of integrating and advancing SRE practices is ongoing, and each step forward enriches our capacity to build stronger, more reliable systems. Through continuous learning and application of these advanced techniques, we can anticipate and mitigate potential failures more effectively, driving innovation in maintaining and enhancing system robustness.

Future Outlook

Industry adoption of AI within SRE continues to deepen, with emerging patterns suggesting that its role is becoming more operationally embedded rather than simply assistive. Observability platforms are increasingly integrating predictive and causal models not just for anomaly detection but to support continuous system understanding—identifying failure patterns, degradation paths, and early risk indicators in complex, distributed environments. At the same time, generative AI is being explored and, in some cases, deployed to automate documentation and procedural content, particularly in high-frequency, low-variance scenarios like post-incident summaries or everyday remediation tasks. Several vendors have begun offering AI-assisted runbook generation or dynamic incident reporting tied directly to telemetry, hinting at a trajectory where operational knowledge is produced as a byproduct of system activity, not after-the-fact documentation. While these applications are still maturing, they suggest a directional shift: toward systems that not only observe and react but also synthesize and formalize. AI synthesizes real-time data into usable insights by automatically correlating telemetry signals, identifying anomalies, and suggesting actionable next steps. For example, it can generate dynamic runbooks or propose remediation actions based on patterns learned from previous incidents, enabling teams to respond faster and with greater confidence. If current adoption patterns continue, SRE teams will likely increasingly rely on AI-generated knowledge assets, such as dynamically maintained runbooks, decision support for SLO tuning, and context-aware incident retrospectives, as a standard part of their workflow. Rather than replacing engineers, the trend points to AI serving as a scalable layer for capturing and operationalizing what teams learn over time, helping reduce cognitive load and accelerate response in high-complexity environments.

Summary

In this concluding chapter, we explored the rapidly evolving landscape of site reliability engineering (SRE) and identified key innovations poised to shape its future. Building on the lessons and case studies from previous chapters, we examined how emerging technologies like **GenAI, extended Berkeley Packet Filter (eBPF)**, and **AIOps** are steering reliability engineering toward greater automation, intelligence, and scalability.

A central theme was the rise of **SRE agents**—both vertical (task-specific) and horizontal (orchestration-focused)—which promise to revolutionize how organizations manage system reliability. Vertical agents specialize in areas such as chaos testing, incident triage, or resource optimization. Meanwhile, horizontal SRE agents coordinate these specialized tools, applying a holistic view of system health that aligns SLOs, error budgets, and business objectives. This agent-based model offers unprecedented scalability for teams that have limited human SRE capacity, supporting continuous improvement and self-healing systems.

We also took a deeper look at **Chaos Engineering's** growing maturity. With technologies like eBPF enabling more precise and efficient fault injection, chaos experiments are becoming more granular and adaptable, extending beyond traditional server shutdowns or network disruptions. The interplay of AI-driven experiments, edge computing, and tighter regulatory compliance underscores Chaos Engineering's expanding relevance across diverse industries—from finance to healthcare.

Finally, we revisited the foundational goal of SRE: to foster a **culture of continuous learning and resilience**. Throughout this book, we have seen how SRE practices unify teams and disciplines around proactive reliability, grounded in actionable metrics and transparent postmortems. As we move forward, innovations like AI-driven incident analysis, predictive tooling, and advanced observability dashboards will only deepen this cultural transformation. By integrating new technologies with the core tenets of SRE—empirical testing, automation, and blameless retrospectives—organizations can anticipate failures, adapt quickly, and, ultimately, deliver more robust and resilient services.

By embracing these advancements and continuing to refine SRE practices, you pave the way for **future-ready reliability**—one in which systems can evolve, self-heal, and drive innovation at scale. This chapter and the book as a whole underscore that SRE is not merely a methodology; it is an ongoing journey to construct, operate, and continuously improve systems built for real-world demands.

Index

A

Accelerated change management, 125, 126

Access control, 89

Agentic AI
- automated triage, 260
- capacity planning, 261
- change management, 261
- chaos engineering, 261
- incident detection & observability, 259
- knowledge management, 262
- root-cause analysis, 258, 260
- self-healing, 260

Agile, 13, 14, 19, 20, 46, 51, 92–97, 162, 175, 181, 182, 205, 273, 297

AI, *see* Artificial intelligence (AI)

AI agents, architecture, 262–264

AIOps, 216, 217, 227, 241, 246, 247, 250, 301
- continuous cycle of hypothesis, 217–219
- use cases, 220–229

Alerting mechanisms, 120

AML, *see* Anti-money laundering (AML)

Anti-money laundering (AML), 24

APM, *see* Application performance monitoring (APM)

Application owner, 94

Application performance monitoring (APM), 221

App Store, 2

Artificial intelligence (AI)
- contemporary implementations, 240
- data to drive decisions, 245, 246
- generative AI (GenAI), 252–259
- measurement, 243
- predictive analysis, 246–251
- risk embracement, 241
- toil management, 243

Audit log, 123

Automation, 3, 63, 67, 76, 118, 122, 126, 136, 148, 196, 211, 240, 244, 261
- and orchestration techniques, 229, 230
- remediation, 246–251
- toil management, 169, 170

Autonomous teams, 208

Autonomy and flexibility, 85

Autoscaling, 261

Azure Chaos Studio, 268

B

Backlog management, 106

BDD, *see* Behavior-Driven Development (BDD)

Behavior-Driven Development (BDD), 94

BigPanda, 249

Blameless postmortems, 141, 142, 262

Blast radius, 268, 275, 276

Bleeding-edge tools, 24

© Florian Hoeppner, Francesco Sbaraglia 2025
F. Hoeppner and F. Sbaraglia, *Mastering Site Reliability Engineering in Enterprise*,
https://doi.org/10.1007/979-8-8688-1448-8

C

GPSR Compliance
The European Union's (EU) General Product Safety Regulation (GPSR) is a set
of rules that requires consumer products to be safe and our obligations to
ensure this.

If you have any concerns about our products, you can contact us on

ProductSafety@springernature.com

In case Publisher is established outside the EU, the EU authorized
representative is:

Springer Nature Customer Service Center GmbH
Europaplatz 3
69115 Heidelberg, Germany